Yonas Berhane

Mezard, indigenous conflict Resolution mechanism in northern Ethiopia

Assessing rural alamata woreda, Tigray regional state, Ethiopia

Anchor Academic
Publishing

Berhane, Yonas: Mezard, indigenous conflict resolution mechanism in northern Ethiopia Assessing rural alamata woreda, Tigray regional state, Ethiopia, Hamburg, Anchor Academic Publishing 2013

Buch-ISBN: 978-3-95489-157-3
PDF-eBook-ISBN: 978-3-95489-657-8
Druck/Herstellung: Anchor Academic Publishing, Hamburg, 2013

Bibliografische Information der Deutschen Nationalbibliothek:
Die Deutsche Nationalbibliothek verzeichnet diese Publikation in der Deutschen Nationalbibliografie; detaillierte bibliografische Daten sind im Internet über http://dnb.d-nb.de abrufbar.

Hermannstal 119k, 22119 Hamburg
http://www.diplomica-verlag.de, Hamburg 2013
Printed in Germany

Acknowledgement

I am grateful to Gebremedhin Yihdego (Ass.prof.), my advisor, for his scholarly enlightening supervision in carrying out the research work. His scholastic insight and the genuine classroom reflection have also influenced me positively in the course he offered me during my study. In line with this I deeply appreciate my co-advisor Abraham Worku (MA) for his constructive comments he gave me throughout the research work.

My warmest and honorable thanks also goes to Abreha Hailezgi, lecturer in Mekelle University, for all the encouragements and his honest and constructive preliminary piece of advice he gave to me regarding the topic of the research.

Moreover, I would like to thank to all my informants without whom the research would have borne no fruit. I especially thank Ato Mesfin Andarge and Ato Rezene Hagos for their cooperation in meeting and convincing the resourceful informants. I also thank, Alexander Berhane, Mentsegeba Girmay and all my friends who have their own shares in the accomplishment the research work.

Thanks to GOD!!!

Yonas

June 2012

Table of Contents

Abstract

This study deals with indigenous conflict resolution mechanism in Rural Alamata woreda specifically in the institution called Mezard exercised by the Raya people. The study was aimed at examining the Mezard indigenous institution of conflict resolution. The research design employed for this study was case study research design. To this end purposive sampling design was used, employing 24 total sample sizes. Its composition was from indigenous judges, court leaders, to explore the sustainability of the institution, the enforceability, strengths and weaknesses and its role in maintaining peace and stability in the area. This study was conducted using unstructured interview as the main data collection instrument along with focus group discussion employing open ended questions to supplement the data gathered from the key informants. Accordingly, the study finding shows sustainability of the institution is being in formidable challenges, this is mainly because of its less attention on transferring the practice to the young generation and some forces of modernization, etc. With regard to enforceability, the decisions passed by the institution are found to be high in being accepted by the community of the study area. In order to integrate Mezard indigenous conflict resolution mechanism with the formal court legalizing their role and defining their jurisdiction is so important. In addition advancing affirmative action, increasing the income of women by creating new job opportunities, give chance the youth to decide on their matters continually, raising finance generated from the community, etc can help the sustainability of the Mezard institution of conflict resolution.

Key words: *Erq, indigenous conflict resolution, Mezard, Tabia, Wonz*

List of Acronyms and Abbreviations

ACCORD	The African Center for the Constructive Resolution of Disputes
ADR	Alternative Dispute Resolution
CSA	Central Statistical Agency
DCG	Dry Land Coordination Group
DFID	Department for International Development [United Kingdom]
ICRM	Indigenous Conflict Resolution Mechanism
IPMS-ILRI	*Alamata* Pilot Learning Site Diagnosis and Program Design, International Livestock Raring Integration
USAID	United States Agency for International Development
FGD	Focus Group Discussion

CHAPTER ONE: INTRODUCTION

1.1. Background of the Study

Conflict is not a new phenomenon in the history of human beings. It always exists as long as human beings live together and it is an inescapable human experience. Conflict can be understood as engagement in a fight or possible confrontation between two or more parties as aspiring towards incompatible or competitive means or ends (University of Peace, 2005). According to Francis (2006), conflict is by definition natural and unavoidable part of human existence, i.e. a nature disagreement resulting from individual or groups that differ in attitude, beliefs, values or needs.

As there is the notion of society and the very existence of peoples, there is obviously conflict no matter the extent as well as the source of it. In fact, one cannot separate conflict from humankind. It is impossible to see them independently; they are always together making two faces of the same coin. It is unconceivable to think a society without conflict (Assefa, 2005).

Conflict appears to be an integral component of human performance in diverse activities. The term could be classified based on different criterias and variables, and this creates difficulty in formulating an operational and exact definition. According to Slabbert (2004), conflict is a process that begins when one party perceives that another party has negatively affected, or is about to negatively affects, something that the first party cares about. Hence, attempting to avoid conflict is not only going against nature but it is also going against development. However, conflict should be wisely resolved before it turns to be destructive.

What especially makes conflict normal and natural is its very existence at different levels and walks of life of people in a given society. In fact, conflict takes different forms to occur. It may come to happen at intrapersonal, interpersonal and inter-group levels. It may also be within and/or between communities (Befekadu and Diribssa, 2005).

The practice of conflict resolution is dominated by the assumption that conflict can at best be managed but hardly possible to solve it. Particularly, conflict management theories see violent conflict as an ineradicable-viewing resolving a conflict as unrealistic and the best that can be done is to manage and contain them and occasionally to reach a historic compromise (Reilly.L and Bloomfield.T, 1998).

On the other hand Conflict Resolution theorists argue that it is possible to transcend conflict if the parties can help to explore, analyze, question and reframe their position and interest.

Conflict resolution, therefore, emphasizes intervention by skilled but power third party to foster new thinking and new relations (Miall, 2003). Thus conflict resolution involves process oriented activities that aims to address the underlying causes of direct, structural and cultural violence (Walters, 2000), while simultaneously finding a way to respect their underlying values and identities (Spangler, 2003).

There are two ways of conflict resolution methods, i.e. the formal and informal. According to Macfarlane (2007), the formal one is conducted in the court through litigation with official recognition of the government and tends to be more strictly examined and monitored whether there are signs of impartiality, entrenched inequities and lack of due process. Indigenous or sometimes called informal conflict resolution mechanism is the ancient set of practices in almost all the societies in the world with an ultimate application of third party arbitration and mediation (United Nations, 2007).

According to Kennedy (2011), the concept of 'indigineity' connotes institutions, mechanisms and practices predating colonialism and the Westphalia state that derive from the sociological, historical, demographic, environmental and geographical contexts in which they exist.

Kennedy, (2011) suggested that the role of indigenous-based institutions and culture in conflict prevention and resolution as well as peace building is becoming increasingly prominent in most countries.

Dahal and Bhatta, (2008) also confirmed that indigenous conflict resolution mechanisms are used to organize a discussion to mediate several types of conflicts within the societies by recognizing the identity and interests of the conflicting parties. Here its main aim is to come up with a solution and justifying it by the duty of the mechanism based on recognized customs and morality of the society itself.

"Indigenous Dispute Resolution processes," are intuitive, time-tested and pre-colonial forms and systems of dealing with community problems by coming up with a consensual, communal solution (Osi, 2008).

'Indigenous Institutions' include structures, mechanisms, methods or techniques and practices that incorporate indigenous principles such as conciliation, discussion, negotiation, mediation and arbitration. Indigenous approaches cannot be compartmentalized into "political" or "juridical" or other, rather they are holistic; comprising also social, economic, cultural and religious-spiritual dimensions

This implies that the sum of indigenous lifestyles and world views in which the different spheres of societal life are hardly separated. In addition the conflict parties can directly engage in negotiations on conflict resolution and in the search for a solution, or a third party can be invited to mediate.

In this case Kennedy, (2011) suggests that the process of conflict resolution is public and the participation in the process and the approval of results is voluntary.

The process is carried out by social groups in the interest of social groups such as extended families, clans, village communities, tribes, brotherhoods, etc.; individuals are perceived as members of a kin-group, they are accountable to that group, and the group is accountable for the deeds of each of its members (Boege, 2006).

The above idea implies, every society has its own culturally deep-rooted mechanisms of conflict resolution. Accordingly, *Raya* is one of the cultural communities in Tigray and northern Wollo, has its own indigenous mechanisms of conflict resolution. Thus, this research paper exhaustively deals with how any sort of conflicts are settled and resolved by the *Mezard* and how its sustainability and enforceability is in the study area.

The *Raya* people have their own varieties of customary practices. Among these practices they have traditional conflict resolution mechanisms which are more than helping the people to build solidarity and peaceful coexistence among them. It has a gap filling role for the state made laws. In Raya, offices of elderly are known by variety of names, but they do have common goal i.e. resolving the conflicts especially personal and family peacefully through mediation, conciliation, and arbitration mechanisms at local level customarily. *Mezard, Zewold, Kefco, Gebremedhin, Mergeta, senayseged,* are some of the most known indigenous mechanisms with local judicial jurisdiction. There are various indigenous institutions in Ethiopia that have their own customary methods of settling conflicts. For the purpose of this paper the researcher dealt with the *Mezard* indigenous conflict resolution mechanism which is practiced in the rural *Alamata* woreda. This institution was chosen for its unique characteristics in form, procedures and content. In addition it is predominantly exercised by the residents of the study area.

1.2. Statement of the Problem

Africans, as any other people elsewhere, have their own philosophy of life. They have a distinct culture by which they see and interact themselves. For instance, individualism is a philosophy of life in the West as communality is in Africa. Collective mode of living is a peculiar identification of Africans in general. In fact, this mode of living is a different version of life rather than connoting backwardness and/or primitiveness which might be wrongly perceived by some westerners. The communal life in Africa, on the other hand, has an implication in conflict and its resolution. The issues of security, peace and conversely conflict are communal matters in Africa, unlike in the West. Thus, the priority of African traditional conflict resolution mechanisms is to bring about collective social stability and harmony. They do not simply focus on the punitive measure nor do they merely settle the conflict. They rather strive to bring about sustaining peace among the disputants thereby the prevalence of enmity and hatred within as well as across communities would permanently be vanished (Tsega, 2002).

Conflicts in Ethiopia could easily be handled at the grass-root levels by the societies themselves without the direct intervention of the government. After all, Ethiopia is believed to be the home of more than 80 ethnic groups. They have their own distinct languages, and cultures. These various ethnic and cultural groups for so long years have developed their own unique political or administrative, economic, social and judicial systems. They had been able to sustain themselves without necessitating to have copied the Western modes of governance nor is their mechanisms of conflict resolution (Assefa, 2005).

Indigenous laws have been "de-facto" recognition in Ethiopia and has been applied both in formal courts and traditional justice processes and in relation to interpersonal and community based conflicts. Under the Constitution, its application is however, restricted to matters of personal law and therefore its application in other matters relating to conflict does not have any legal recognition and enforcement, as the indigenous institution of conflict resolution is subordinate to the formal court (ibid,2006).

Several researches are done on several indigenous conflict resolution mechanisms in Ethiopia in among nations, nationalities and ethnic groups of Ethiopia. These among others include such as Gereb in Wajerat (Tigray) and Mablo in Ab'ala of Afar (Kelemework, 2000); Abagar in Southern Wollo (Netsanet, 2006), Michu in Metekel (Tsega, 2002), Awassia in Walayta (Mellesse, 2008), Afocha in Harar (Biruk and Jira, 2008), Jarsumma and Haffi Hamee in Oromia (Assefa, 2005), Guma in Waliso Oromo (Dejene, 2007) and so on and so forth.

However the above researches were predominantly relied on the process, role of the elders, as well as the actors in the indigenous conflict resolution mechanism. However, this study is intended to search the issues dealing with the sustainability and the enforceability of the decisions passed by the indigenous conflict resolution which is not well addressed by the previous researches.

Similarly, the Raya community living in the southern part of the regional state of *Tigray* in *Alamata woreda* practice their own way of conflict settlement. Customarily they have their own indigenous way of conflict resolution mechanisms aiming to address the cause of conflict and seek to build solidarity and belongingness for themselves and for peaceful and good relationship within themselves and their neighboring societies.

In recent years the indigenous conflict resolution mechanisms have been weakened partly because of the failure of the members of the community to strengthen their role of resolution of conflicts at the community level (Asnake, 2010). Therefore, the undermined and eroded status and functions of the elders give rise to the absence of well performing institutions which is a sufficient condition for violent conflicts to arise.

The conflict resolution mechanism called *Mezard* plays a complementary role in keeping the peace and stability among the people in the study area. Despite the role it plays and legal recognition it has, the government of the state gives less regard to it. Members of this indigenous tribunal perform their duty permanently at least two days a month but without any allowance and budget from both the people as well as the administration for the activity they conduct. As a result the elder traditional tribunals sometimes become less sensitive to the cases brought before them. This condition leads them to some irregularities in their activity among themselves and the conflicting parties. Therefore the general purpose of this study is to assess how the Mezard institution of conflict resolution mechanism handles any sort of conflict in the study area and how is its current status.

1.3. Research Questions

The study addresses the following research questions:

- What are the types and sources of conflict in the area handled by the *Mezard indigenous* conflict resolution mechanism?
- How is the procedure of *Mezard* conflict resolution mechanism practiced in the rural *Alamata* woreda?
- What is the enforceability of the *Mezard* indigenous conflict resolution mechanism?
- How is the sustainability of the *Mezard* indigenous conflict resolution mechanism?
- What are the strengths and weaknesses of the *Mezard* indigenous conflict resolution mechanism?

1.4. Objective of the Study

1.4.1. General Objectives

- The general objective of the study is to assess the *Mezard* indigenous conflict resolution mechanism in rural *Alamata woreda* Tigray region of Ethiopia.

1.4.2. Specific Objectives

- Identify the types and sources of conflicts in the area with regard to the *Mezard* conflict resolution mechanism.
- Describe the procedure o*f Mezard* as conflict resolution mechanism in the Rura*l Alamata woreda.*
- Examine the enforceability of the *Mezard* indigenous conflict resolution mechanism.
- Examine the sustainability of the *Mezard* indigenous conflict resolution mechanism.
- Examine the strengths and weaknesses of *Mezard* indigenous conflict resolution mechanism.

1.5. Significance of the Study

Conflicts occur among individuals, within families, between clans, and communities across ethnic borders. The study shows indigenous conflict resolution method's (ICRM's) relevance in settlement of conflict especially as family, inter-ethnic, etc across regions. Hence indigenous laws are made by the people and not the state and also drive its legitimacy from the participation and consensus of the community and its recognition by the government. This study has the following relevance;

- Serving as an input for policy makers in relation to conflict resolution and peace building.
- Serving as a reference for other researchers to conduct research on similar topic region-wide or nation-wide.
- Contribute to a better understanding about the indigenous conflict resolution method '*Mezard*' to the community.
- Help to specify the factors that hinder the indigenous conflict resolution methods from application and contribute for further strengthening the institution to the extent that it go line in line with the formal court.
- Additional contribution to the few works that have already been done on conflict resolution in the region. It serves as a useful input for those individuals interested to carry out in-depth studies on similar topics.

1.6. Scope and Limitation of the Study

The Raya people have a number of indigenous practices with socio-cultural and economic practices. The study is only limited to Tigray region in general and *Raya Alamata* woreda in particular on the indigenous mechanism of conflict resolution in rural Alamata. Comprehensive evidence can be obtained only by working within the institution or taking a longitudinal research method. This will be one of the methodological limitations of this research. In addition to this the research only employs a convenience sampling procedure.

1.7. Definition of Concepts

***Abahaga*:** the one who is in charge and of something; care taker of one social issue

Anebabro: bread like food

***Duberti*:** a group of women in the community who beg the conflicting parties to come in to negotiation then to arbitration

***Erq(i)*:** resolution or reconciliation.

***Ketena*:** an administrative structure found below woreda and above the *tabia*

Mak: a black thick cloth

***Mereba*:** a traditional request made by the elder women to beg *erq*

***Meyazna*:** an agent as a guarantee

Mezard: is an indigenous or informal way or method of conflict resolution which is practiced around the rural Alamata woreda in southern part of Tigray region also kefko, gebremedhin, mergeta, senayseged, zewold

Millisha: one who is armed from the members of the community, mostly doing his civil matters too

***Raya*:** is a name specifically given to both the area which is found on southern Tigray and Wollo as well as the people who reside in those areas

Rond: night guard

***Tabia*:** is one of the Government structures within the regional states which is smaller in unit as well as less powerful than the woreda

***Teskar*:** ceremony conducted for commemoration of the deceased

***Wanza*:** an acacia tree

Wonz: literally in Amharic means house of elder fathers

***Woreda*:** is an administrative structure which is equated with the District

Yerejat: close relatives of the murderer

CHAPTER TWO: REVIEW OF RELATED LITERATURE

This chapter exhaustively deals with the theoretical framework of the nature and essence of conflict and types of conflict, conflict resolution mechanisms with special reference to the indigenous dispute resolution. Only some of the most prominent African traditional institutions in general and the Ethiopian in particular which are currently being practiced and used in resolving conflicts of any sort are also pointed out and reviewed from different sources.

2.1. The Concept of Conflict

The term conflict refers to a situation of hostility between individuals, groups, or states resulting from competition over resources, power, or opposing interests in other areas (Francis, 2006).

Individuals need each other for survival; along the way conflict is unavoidable. What is important for all parties is that differences be reconciled without resort to violence. Goodhand and Hulme (1999), note that 'conflict is only a problem when society cannot represent, manage or resolve its different interests in a productive manner, thus initiating a degenerative or destructive cycle of physical violence.

Conflicts are as old as human societies themselves. Historically, individuals, social groups and societies have disputed and competed against one another over scarce commodities and resources land, money, political power, and ideology. They have even fought one another and bitterly sought the elimination and/or subjugation of rivals, in order to control these resources and commodities (Reichel, 1998). Similarly conflict is defined as a situation in which two or more parties strive to acquire the same scarce resources at the same time. It occurs between people in all kinds of human relationships and in all social settings. Because of the wide range of potential differences among people, the absence of conflict usually signals the absence of meaningful interaction. Conflict by itself is neither good nor bad. However, the manner in which conflict is handled determines whether it is constructive or destructive (Deutsch & Coleman, 2000).

In addition, Conflict refers when two or more values, perspectives and opinions are contradictory in nature and have not been values and perspectives are threatened; or discomfort from fear of the unknown or from lack of fulfillment. Conflict is inevitable and often good, for example, good teams always go through a "form, storm, norm and perform" period. Getting the most out of diversity means often contradictory values, perspectives and opinions (McNamara, 2007).

Nathan (2007) on the other hand asserted that conflict is also a nature consequence of major reform and of popular pressure for fundamental political or economic change. Furthermore, (Spangler,2003) stated that conflict tend to arise over non-negotiable issues such as fundamental human needs, intolerable moral difference or high stakes distributional issues regarding essential resources such as water or land. Accordingly, conflict is inherent to all societies. Differences in interests and opinions between groups are natural, but the method by which such differences are expressed and managed determine if conflicts manifest themselves in either non violent or violent ways.

2.1.1. Types of Conflict

Conflict exists mainly in two forms. There are direct and indirect depending on their magnitude and manifestation. Direct conflict is the most outward, visible and easily identifiable sort of conflict. It is just like a direct physical confrontations or clashes between individuals. It could also be manifested in a form of war between nations or groups of different interests. On the other hand, indirect violence, which is synonymously referred to as structural violence, appears to be in a form of invisible and relatively unidentifiable societal structures. Such violence is built deep in to the social, economic and political structures. These structures deliberately prohibit every people from enjoying equal social status and equal access to economic opportunities as well as involvement in one's own political matters. It is all about discrimination and marginalization, suppression and exploitative world order. And all these in turn would result in human suffering and life misery in general as the consequence of impoverishment to overwhelmingly majority of the people (Jeong, 2005).

Hence, violence or conflict of any kind needs to be peacefully managed and resolved. If not so, it would result in human destruction. There might be constructive conflict but only if it is properly managed and harnessed as to be supporting and driving force for humans in their development.

Macfarlane (2007), outlines that, there are two forms of conflict management and resolution mechanisms as formal and informal conflict management and resolution that the best result often are achieved by combining the two methods.

The formal way of conflict management and resolution is conducted through litigation as the conflicting parties are subject to court procedures and final judgment thereby the parties become binding to the court decisions, where as the informal one is accompanied with major components such as mediation, arbitration, and negotiation among the others (Brainch, 2006).

2.2. Indigenous Conflict Resolution Mechanism

As long as people live in society or group, there are conflicts arising from differences of interests, prejudice, needs and ambitions. Therefore, the approach adopted to prevent or resolve such difference of interests determines its resolution. In other words, when a conflict happens, the crucial point should be the effective adoption of the necessary principle of the resolution. Indigenous mechanisms are grass root approaches to solve conflicts by the society. The most important elements involving in this mechanism include the tradition of forgiveness, respect for elders because of their symbolic authority to enforce decisions and transfer of resource as compensation (Zartman, 2000).

Indigenous conflict resolution typically incorporates consensus-building based on open discussions to exchange information and clarify issues. Conflicting parties are more likely to accept guidance from these mediators than from other sources because an elder's decision does not entail any loss of face and is backed by social pressure. The end result is, ideally, a sense of unity, shared involvement and responsibility, and dialogue among groups otherwise in conflict (USAID, 2005).

Some authors trace the genealogy of what is now known as mediation to early Greek and Chinese cultures. Sri Lankan antecedence dates back to about 425 years before Christianity while Chinese mediation goes back to the Zhou Dynasty which predates the Christian era by 1,100 years. Other scholars characterize pre-colonial forms of mediation as a familiar face of early Jewish, Christian, Islamic, Hindu, Buddhist and Confucian religions as well as Indigenous cultures. Ethnic groups and aboriginal tribe's people have been practicing these forms of mediation long before newer settlers arrived (Osi, 2008).

Indigenous conflict management and resolution mechanisms use local actors and traditional community-based judicial and legal decision-making mechanisms to manage and resolve conflicts within or between communities. Local mechanisms aim to resolve conflicts without resorting to state-run judicial systems, police, or other external structures. Local negotiations can lead to ad hoc practical agreements which keep broader inter-communal relations positive, creating environments where nomads can graze together, townspeople can live together, and merchants can trade together even if military men remain un-reconciled (Lowry, 1995).

While Western forms of ADR (alternative dispute resolution) were generated as a response to the difficulties and deficiencies associated with court proceedings, Indigenous Dispute Resolution processes were not an “alternative” to anything. There were no courts or highly formalized procedures and institutions to speak of when they were first developed and practiced. Generally, Indigenous Dispute Resolution processes were all that the communities had in common. Although they seem very ADR-like, they were truly indigenous and unique to these peoples (Brainch, 2006).

Indigenous societies have varied institutions and personnel that function as agents of conflict resolution in their respective locality (Degene, 2007). In spite of this, Ember (1977), cited in Degene (2007), argued that, these agents lack coercive force to support their decisions.

There is a little evidences from the literatures that a settlement of conflict within the indigenous institutions is in any way encouraged by the governments, rather the indigenous institutions of conflict resolution impose sanctions which are of two types, the secular and the sacred from which the later is more effective and often used(Hamer,1987).

Indigenous societies in all parts of the world have featured variations of third-party arbitration and mediation. Western societies saw these practices subsumed by the rise of modern judiciaries. The increased complexity of these processes, however, saw reduced satisfaction with legal outcomes among disputants (United Nations, 2007).

In fact, the indigenous forms of conflict resolution have been practiced by peoples and communities for centuries. The older forms of dispute resolution, particularly those practiced by the Indigenous or Aboriginal peoples around the world, challenge the originality of present-day court system (Osi, 2008).

Indigenous conflict resolution offers not only alternative means of resolving conflicts but an entirely more efficient way of doing business with each other. It’s the term in advance of justice all over the community but remains widely misunderstood which is likely to be the reason it is often initially treated with less recognition by the state (Brainch, 2006). According to Open University of Nigeria (2010), these institutions were mostly characterized by public will and associational trust especially as they are conducted truthfully and pragmatically.

Indigenous conflict resolution comprises a range of processes fashioned structured to meet the specific needs of parties in a conflict; each process being an alternative to litigation but commonly 'mediation' and 'arbitration'. The key principle is that the means used in seeking to resolve the conflict must be proportionate to its nature in terms of value, complexity and importance to the conflicting parties (ibid,2006). This implies that Indigenous conflict resolution mechanisms refers the set of mechanisms a society utilizes to resolve conflicts outside litigation or the formal court (Osi, 2008).

Indigenous or Aboriginal communities in the present day practice three forms of conflict resolution their own Indigenous processes, Western forms of ADR, and the Indigenized Western ADR systems. They sometimes resort to court litigation, whether because it is expensive, time-consuming, highly complicated, and judge and counsel-dependent, or simply because it is not innate for them to do so. Of course, certain exceptions have to be made for a few communities (Brainch, 2006)

2.2.1. Advantages and Disadvantages of Indigenous Conflict Resolution Mechanisms.

Some of the advantages of indigenous conflict resolution, in contrast with the court litigation or formal way of conflict resolution are given as it has low cost, speed, accessibility, cultural relevance, and responsiveness to the poor people's concerns(DFDI,2004). According to penal reform international (2003), also indigenous is advantageous as the conflicting parties are active participants, raising public interest and awareness, gives rise to permanent resolution of conflict as well as it employed non discriminatory approach on its conduct.

Indigenous conflict management and resolution mechanisms aim to resolve conflicts locally, preceding or replacing external conflict resolution and thereby reducing reliance on external structures. Indigenous mediation helps the community keep control over the outcome of the conflict. Implementing this approach does not require sophisticated party structures or expensive campaigns; it provides a low-cost, empowering means of resolving conflicts within a relatively short time frame. In many societies, elders have indigenous jurisdiction in facilitation, arbitration, and monitoring outcomes (Paul, 1994).

Indigenous conflict mediators typically possess moral status, seniority, neutrality and respect of the community; they are acceptable to all parties and demonstrate leadership capacity. Resolutions are generally accepted and respected by all concerned parties (USAID, 2005).

According to Lowery (1995), some indigenous conflict resolution efforts may be weakened by age or gender bias for example, in cases with no women elders, some women may believe that male elders are biased against women and that this will be reflected in their decisions. Therefore the above idea can imply that the Indigenous authorities are predominantly not progressive elements of social change.

International agencies' efforts to build local capacity and enhance participation should question whether indigenous structures are being undermined, what their role is in keeping the society intact and managing conflict, and whether it is important to make efforts to retain such structures. Indigenous mediation has a dynamic of its own and does not always respond positively to external prompting. Indigenous mediation requires delicate and knowledgeable management, and external actors must bring an intimate understanding of local conditions (paul, 1994).

Indigenous conflict resolution processes will allow community members to pursue remedies and resolve conflicts outside of the courtroom and still within their own cultural confines. Practiced in Indigenous communities since time immemorial, they are culturally more appropriate than litigation because they are based on the customs and traditions of the group concerned. They may also pursue remedies through state-formalized ADR. Litigating in court is normally prohibitive; with long case queues, intermittent delay in the resolution of motions or claims is quite common. More importantly, litigation with its basic rudiments of confrontation, fault-finding and judge made resolutions, coupled with its adversarial nature (Osi, 2008).

Assefa (2005), has summarized only some of the advantages of indigenous conflict resolution mechanisms as they quickly respond to crisis in terms of time, they contribute to the reduction of regular court case loads, they contribute to saving of the public money, it also minimizes the problem in shortage of judges who work in the regular courts, and budget constraints, they are complementary to modern government structures and are not substitutes or competitors as some government officials think and worry about, They give access to many people who do not find modern system of conflict resolution comfortable, affordable or suitable to their needs, Disputants are satisfied with their operation and view their outcomes as fair, and the like.

According to DFID (2004), the disadvantages of indigenous conflict resolution mechanisms are among the others, it may led to the abuse of power, may lack accountability, and not consistent with International human rights as it practices inhuman and degraded punishments. The center for Democracy and Governance (1998), also suggested that, indigenous conflict resolution mechanism has limitations such that it is not appropriate taking cases as public sanction, with extreme power imbalance, may undermine other judicial reform efforts, and have no any educational, disciplinary, or restriction effect on population. Therefore the decisions that can best describe court proceedings are adversarial, thorough, rigid, strict, evidentiary, rights-oriented, unbiased, impartial, and generally appealable unlike the indigenous conflict resolution mechanisms. On the other hand Boege (2006), noted that, indigenous conflict resolution is targeted at problems in relatively small communities in the local context. It can work well within a given community with regard to the members of that community. Conflicts within and between families, between neighbors, within and between villages or clans lend themselves rather easily to traditional approaches.

There are several problems with using Indigenous Dispute Resolution for more advanced cases such as patent challenges concerning Indigenous communities.

Firstly, Indigenous populations have different fundamental conceptions from mainstream society, resulting in only a handful of people who actually understand the culture indigenous to them. This condition can lead to misunderstanding, mistrust, abuses and marginalization. Secondly, Governments may hesitate to organize or uphold Indigenous Dispute Resolution infrastructures, whether in traditional conflict areas such as land disputes and employment discrimination to newer conflict prone subjects such as patenting of traditional knowledge, nonconsensual commercial use of indigenous symbols or art, and the unauthorized recording/reproduction of ancient chants, songs, movements and dances (Osi, 2008).

Accordingly Boege (2006), identifies the following potential advantages and disadvantages of the indigenous conflict resolution mechanisms as follows.

Strengths of traditional approaches to conflict transformation can be identified since the indigenous conflict resolution approaches;

- *fit situations of state fragility and failure;*
- *are not state-centric and hence credited with legitimacy;*
- *take the time factor into due account and are process-oriented;*
- *provide for comprehensive inclusion and participation;*
- *Focus on the psycho-social and spiritual dimension of conflict transformation.*

The five major strengths of indigenous approaches to conflict transformation are matched by five major weaknesses. Indigenous approaches;

- *do not terminate violence in the long term;*
- *often contradict universal standards of human rights;*
- *have a limited sphere of applicability;*
- *are geared towards the preservation of the 'good old' order;*
- *are open to abuse(*Boege, 2006, pp.22-27)

2.2.2. Actors in Indigenous Conflict Resolution Mechanisms

The process is led by the male or female leaders of communities, such as traditional kings, chiefs, priests, healers, big men, elders "elder" being a social, not a biological category and others. These authorities are the mediators, facilitators, negotiators, peace makers. They are highly esteemed for their knowledge of custom, myths and the history of the communities and the relationships of the parties in conflict (Hamer, 1980).

Indigenous conflict resolution's rich experience in conflict regulation, their skills in setting and interpreting signs of reconciliation and their skills as orators as well as their social capital as leaders of the communities empower them to negotiate a resolution to the conflict that is acceptable to all sides. The leaders and other third parties have no or hardly any sanctions at their disposal that would allow for the enforcement of a settlement (Zartmman, 2000).

There is no monopoly over the legitimate use of violence as in modern states that would make it possible to force parties to abide by the law or to accept a ruling. Sanctions are confined to the social realm as shaming, stigmatizing and the supernatural as cursing, sorcery if one does not want to take recourse to violence. Parties have the power to reject any settlement that they are not happy with. Hence conflict solution is based on voluntary consensus and agreement. Everybody has to agree to a solution, including gods and the spirits of the ancestors. In other words, indigenous conflict transformation is consensus-based (Kennedy, 2011).

Community members involved in the conflict participate in the dispute resolution process. These community members can include traditional authorities as elders, chief's women's organizations, local institutions and professional associations. The elders function as a court with broad and flexible powers to interpret evidence, impose judgments, and manage the process of reconciliation. The mediator leads and channels discussion of the problem (Boege, 2006).

When solutions have been achieved, they are sealed in highly ritual forms. Ceremonies are of great symbolic and practical importance. They are a means of conflict transformation in their own right. The whole community participates in them. They are loaded with spiritual meaning. Wealth exchange, prayers and sacrificing to god/the gods/the spirits of the ancestors are part of those ceremonies as well as customary rituals such as breaking spears and arrows, drinking and eating together, singing and dancing together or consuming certain drugs (Osi, 2008).

Parties typically do not address each other, eliminating direct confrontation. Interruptions are not allowed while parties state their case. Statements are followed by open deliberation which may integrate listening to and cross-examining witnesses, the free expression of grievances, caucusing with both groups, reliance on circumstantial evidence, visiting dispute scenes, seeking opinions and views of neighbors, reviewing past cases, holding private consultations, and considering solutions (DFID, 2004).

2.2.3. Sustainability of Indigenous Conflict Resolution Mechanism

The future of indigenous conflict resolution mechanism can be seen as hopefully. Because many states began to adopt them in line with the formal court litigation. As it is outlined by DFID briefings some of the country experiences around the world are given below;

- *Shalish, Bangladesh: takes the form of a public event in which civil disputes are resolved through arbitration and/or mediation, by people with some standing in the community.*
- *Rondas Campesinas (Night Watch patrols), Peru: community-based organizations in charge to settle crime such as theft, slander, assault and domestic disputes they have both policing and judicial functions.*
- *Chiefs and Assistant Chiefs, Eastern Kenya: elected by the government, significant role where access to police and courts is restricted and control*

and record events of cases as elders chosen by the disputing parties make the final decision.

- *Street committees, South Africa: aim to achieve settlement over reprisal. Security mechanisms are also used to ensure personal safety at home and in the community for example, the protection of battered women (*DFID, 2004 pp47-50).

However in contrast with the above idea indigenous conflict resolution mechanisms faces a lot of problems which affect their future sustainability. This is shown with the existing diminishing role of indigenous institutions for settlement of conflicts.

Closely related to the above issue is the fact that indigenous institutions of resolving conflicts among many communities are diminishing from time to time as a result of marginalization by formal dispute resolution mechanisms and civilization. As such, elders are increasingly finding it difficult to prevent and manage conflicts partly because their actions are not secured in any legal framework and also because they lack a coercive mechanism to enforce their resolutions. However, the institution of elders' posses a potential in preventing and managing conflicts especially among pastoral communities provided it's strengthened and backed by legal and policy foundations (Desalegn, et al, 2004).

Despite the imposition by the government and internal changes, many of the indigenous conflict resolution system have retained their basic structure for centuries. The majority of conflicts among the community are still handled at local level. Informal mediators who do not have connection with the modern governmental legal institutions deal with several cases including homicide, which is the gravest of crimes. In addition to local mechanisms, considerable members of a community also opt to government court for legal redress when need be (Degene, 2007).

This condition gives rise to the indigenous conflict resolution mechanisms to get less recognition from any formal institution. Despite its recognition by the majority of the members of the community, in their role in regulating interpersonal relations it is predominantly losing relevance due to the overall, state-imposed rules and regulations (Desalegn, et al, 2004).

Similarly Meron (2010), pointed out that the Ethiopian constitution has, however, limited the mandate of the customary and religious institutions to private and family matters. Nevertheless, these institutions are playing a very significant role in other domains such as criminal matters. The strong social tie existing in the community makes the significance of reconciliation, the key role of indigenous institutions, indispensable.

It is underscored that the state and the NGOs show a strong commitment to working with indigenous institutions as a means of achieving development. However, no pragmatic collaboration is being realized between the statutory and the customary institutions (Huyse, 2008).

Indigenously evolved approaches to consultation, mediation, negotiation and conflict resolution can be effective if social practices, local interest and knowledge are compatible with each other and their synergies contribute to make local conflict resolution mechanisms cost-effective, direct and inclusive of various interest groups. But, it is difficult to see how home-grown approaches have induced desirable change to respond to new conflict dynamics at a time when perpetuation of social and legal order is considered inimical to democratic freedoms, interests and empowerment (Dahal and Bhatta, 2008).

Failure of Existing mechanisms of dispute resolution such as the courts has failed to provide effective administration of justice with respect to certain conflicts such as inter-ethnic conflicts. Therefore, non-recognition of indigenous processes of conflict resolution under law has resulted in there being no clear and comprehensive system of justice to resolve conflicts. This has resulted in disputing parties relying on self help mechanisms that lead to violence and degeneration of relations between parties and their communities (Pkayla and Adan, 2004).

2.2.4. Enforceability of the Indigenous Conflict Resolution Mechanism

Indigenous conflict resolution's potential effectiveness is diminished where indigenous authority has eroded and the formal authority has increased, since these trends run counter to indigenous values and ways of social organization, including those of handling conflict (lowery,1995).

Enforcement of informal decisions presents some challenges on ground since enforcement actions occur when the terms of an agreement have been violated, and thus the institution employ coercive force including seizure of property. However, these mechanisms do have a considerable means of social coercion, which may be more effective than the courts. If parties need an informal decision to be enforced, they can either return to the informal sector or avail themselves of the courts. If the aggrieved party goes to the court, the court would only be able to enforce the decision if the parties had previously sought the approval of the court for the settlement (Pkayla and Adan, 2004).

Resolutions by local mediators may take the form of negotiation or arbitration and are generally reached with reference to the community's norms and values. In theory, proceedings of mediation may be subject to manipulation by the different parties, including the arbitrators. In practice, however, such manipulations seem severely constrained by the elaborate and careful provisions of the community's norms and prescriptions. As soon as guilt is proven, the indigenous law provides guidelines for the amount of compensation or fine to be paid (Yayneshet and Kelemework, 2004).

Documentation on the effectiveness of grassroots conflict prevention mechanisms is inconsistent, yet indicates that indigenous mediation may be powerless to address some of a conflict's root causes centrally-instigated conflict, predatory behavior linked to exploiting economic advantage, external meddling. Indigenous mediators often bring important social influence but may lack the power and the means to enforce the resolutions adopted. Advice is only accepted when both parties agree to it, and both parties must feel their concerns were properly addressed. Traditional structures' power to prevent the occurrence of violence is limited (USAID, 2005).

Indigenous institutions do not impose or enforce outcomes. The only means of coercion available to the community would be shunning or exiling a party that refused to compromise. Thus its major aim is to restore community harmony, which is generally achieved by arriving at an equitable settlement that corrects harm done to honor and/or property (Huyse , 2008).

This system thus works best in small communities and can sometimes be extended to different communities within an ethnic group. It is most problematic when disputes arise between communities of different ethnic groups, particularly if they have an antagonistic relationship. The body of customary law varies from place to place, and impartial mediators may be impossible to identify. Given these weaknesses, failures to resolve serious problems, particularly those involving threats of bloodshed, have historically prompted state intervention to prevent disorder (Blench, et al, 2006).

Pospisil (1990), proposes if the sanctions made by the indigenous institutions really more important and binding, appealing to functional approach that what actually qualifies sanction is its effective social maintenance. He further discusses that though psychological sanctions such as ostracism, ridicule, avoidance or denials of favors are, sometimes very subtle and informal, they are effective sanctions.

Indigenous conflict resolution is rooted in the respect that parties have for the authority and opinion of the community. Only if the disputants see themselves as part of a common community can this form of dispute resolution be effective and its rulings binding. Because communities must first get the disputants to agree voluntarily to mediation or arbitration, the legitimacy of the outcome is determined at the outset when the parties agree to take part and be bound by the results of the process (Barfield and Nojumi , 2004).

This implies that, the community has its own goal of not only coming to a solution for the issues at hand but of reconciling the parties and ending divisions that may generate further conflicts. The closer the community bonds the greater legitimacy of the process. While enhancing effectiveness and legitimacy, this requirement for social cohesion also limits the viability of customary mechanisms in the current environment.

2.3. African Indigenous Institutions of Conflict Resolution

The last decade has witnessed an increasing resort in Africa to indigenous mechanisms of justice, peace, and reconciliation. For example, in Burundi, Kenya, Rwanda, Somaliland, northern Uganda, and several other countries, indigenous mechanisms have gained in appeal as possible alternatives or supplements to approaches to conflict resolution that are largely Western designed. To some observers, these approaches may on the surface appear to be a desperate return to the past optimism of the 'traditional', or nostalgia for the good old days, when things worked. To others they represent a community's realization that the way out and forward lies within it, in its choices, knowledge, wisdom, and understanding of its experiences of war and peace Criticisms of this romantic view merit attention, especially when they warn against the uncritical use of traditional mechanisms blind to changes that may have occurred during the colonial and postcolonial eras. Nevertheless, these criticisms do not necessarily offer a conclusive answer to the debate on whether indigenous ideas of the social have value in promoting understanding of war and peace in Africa. The second view is the point of departure here addressed through a focus on the concepts of flexible gender, *ubuntu*, and design for exploring the significant advantages that indigenous mechanisms offer in understanding war and peace in Africa (Mensah, 2002).

To this end, African societies unanimously have their own cultures of civilizations. It is believed that Africa is the cradle of mankind. As it might also be witnessed in many archeological excavations, there is nowadays a tendency to believe that the first men had probably lived in Africa. Indeed, Africans, at least like any other people elsewhere, certainly have for long time developed their own unique system of administration and governance. As part of their strong and viable system, they had also, and still remain to have, effective and practically workable conflict resolution mechanisms which sustained and solidified them together within their respective communities (Francis, 2006).

This is particularly true of Africa before colonialism ravaged it along with the destruction and erosion of its norms and precious inner most values all over the continent. Western colonizers have not only exploited the African rich natural and human resources but they have also corrupted and poisoned its long time and deep-rooted systems of self governance. It is a common knowledge that Africa had existed with its unique civilization and thereby philosophy long before the intrusion of colonialism to generally disrupt the indigenous knowledge systems (Singleton and Shingler, 1987).

In spite of the fact that traditional societies in Africa are reputed to settle conflicts long lastingly and for that matter to handle their administrative cases in their own, the European colonial powers rhetorically justify colonialism to have a civilizing mission. In fact, colonialism did not only downgrade the indigenous methods of peacemaking but it also had virtually destroyed the democratic institutions in many African societies (Bahru, 2002).

Colonizers did this as part of their purpose to homogenize the world in such away as their values and perspectives would be dominantly flourishing at the expense of vanishing others. In doing so, they replaced the African indigenous practices and institutions which for centuries had upheld the societies by their own imported customs of rule. It is because this would ultimately serve their colonial interests (Mensah, 2002).

In a similar fashion, Wilson-Fall (2000), has also written about how effective the traditional institutions to manage conflicts were in the pre-colonial Africa. According to him and other authors, however, these were intentionally degraded and paralyzed by the colonial governments.

Withstanding all the influences and attempts of destruction, there are still traditional institutions in Africa many of which are coordinately working with the imported modern institutions. In spite of the effect of modernization, the existing indigenous institutions are yet serving to keep harmonies and solidarities among people of a certain society (Zartman, 2000).

Although sufficient attention is not given to the indigenous institutions even in the post-independence era, however, there are to a little extent tendencies to incline to use them applying in the contemporary issues. An example in this regard is the application of gacaca to resolve "modern" conflicts in Rwanda. The Rwandan societies have been using gacaca customarily for so long years at a grass root level to handle any sort of conflicts and/or disputes. The Rwandan government has also recently been able to apply it in dealing with the most deadly and worst crisis of 1994 which had eventually led to genocide reportedly claimed near to a million people(ibid, 2000).

Tongeren et al. (2005) on the other hand, states gacaca as it meet not only sanction over the violators of norms but also the violators are accepted as fully members of the community. That is the reconciliation process is the core objective of the gacaca system of conflict resolution.

Another African traditional institution of this kind, to add, is ubuntu which is also an effective conception of peace making institution. The role of ubuntu in conflict resolution and its very notion is reconciliation and peaceful co-existence based on the principle of reciprocity, inclusivity and a sense of shared destiny between peoples and communities (Ibeanu, 2006).

There is, therefore, a need to promptly revitalize the African indigenous conflict resolution institutions so that conflicts for a long-term are resolved.

Ethiopia as one of the African states is believed to be the "museum of peoples" with more than 80 ethnic groups constituting and forming one nation state. It is a country of diversity harboring varieties of languages, different religions and faiths as well as quite many nations and nationalities with their own philosophical perspectives and unique cultural practices. History also tells us that Ethiopia is as ancient as 3000 years back. It is, therefore, definitely true that it had experienced strong and cohesive systems of administration with its world top leading civilization, namely Axumite civilization. Indeed, every Ethiopian ethnic groups as their African counterparts have customarily age-old and time-tested administrative and conflict resolution institutions at the grass root levels. In fact, exhaustive and thorough studies about these institutions are not bulky enough and so are limited literatures in this area. Be this as it may, however, there are some literatures on only few cases elaborating the importance and strength of these multipurpose institutions. For instance, Bahru (2002), has written about the indigenous political and judicial institutions in Gurage. He pointed out the Yajoba Qicha and Gordanna sera practiced in the society. These institutions are responsible for legislative as well as judiciary functions.

To add one more indigenous institutions of Ethiopia, there is also another such similar multipurpose institution of the Kambata, South nations, nationalities and peoples region. This indigenous institution as to Yacob (2006), is known as seera. Seera is everything to the kambata society. It is everything such that it serves as a basis for the political administration, social interaction and means of conflict resolution within the territory of the society.

Elders in this regard, according to the rule of seera, are considered to have the most esteemed and graceful status. It is believed that it is with the lifelong experiences and accumulated knowledge of the elders which seera would appropriately function (Yacob, 2002). Gada is also the most known political, social and judicial indigenous institution widely practiced among the Oromo people. Much is said and written in different sources about the democratic nature of the Gada system. Hamdesa (2000), and Assefa (2005), also have written that it is an effective institutional system to manage any source of conflicts among the Oromo so that harmony, peaceful co-existence and smooth relations would be maintained in the society. As to Hamdesa (2000), there are thirteen steps on the way to the whole process of conflict resolution through the use of the Gada system to finally arrive at reconciliation.

CHAPTER THREE: RESEARCH METHODOLOGY

3.1. Site Selection and Description of the Study Area

The study was undertaken in southern part of the Tigray National Regional State, in rural *Alamata* woreda. Rural *Alamata* woreda is selected for the study purposively because of there exist a continuous conflict mainly due to natural resources usage. The area is also characterized with various indigenous conflict resolution mechanisms from which the *Mezard* is with unique character in its form, content and procedure. In this woreda the indigenous conflict resolution mechanisms are highly respected among the community and are highly functional. *Alamata* woreda is one of the five woredas in southern zone of Tigray located 600 km north of Addis Ababa and about 180 Km south of the Tigray Regional capital Mekelle. *Alamata* is the south most woreda of the Tigray Region and borders with Amhara region from the south and west and Afar region from the east. There are 15 *Tabias* and two towns in the woreda i.e. *Alamata* and *waja*. The number of the total population of the rural *Alamata* woreda is 20,597.The total population of the Alamata woreda is 85,359 (CSA, 2007). Altitude in the area ranges from 1178m to 3148 m and 75% of the woreda is lowland (<1500 masl) and only 25% is found in intermediate highlands (between 1500 and 3148 masl). The *Alamata* valley is one of the most agriculturally potential areas in the Tigray Region. Farmers in the Woreda extensively cultivate cereals and vegetable; and raise mainly sheep and cattle in the valley. The total area of the Woreda is estimated about 550km2 (IPMS-ILRI, 2005).

3.2. Data Type and Source

This study employs qualitative data type. Both primary and secondary data sources are used for accomplishment of this study. Information gathered through interview with elders and court officials. Focus group discussion is also conducted with individuals who have access to one or more cases of conflict handled by the *Mezard*, youths, and women. In addition to this field observation is conducted as primary data. The people's socio-economic activities, the way they handle the conflicts (process of *erk*) is as such observed in an open or visible manner. Reviewing of different written materials such as books, journals, also enables to have a literature review work and serve to critically examine the problem in the area. Moreover, any compiled documents of the area helps to know the social, economic, physical as well as historical background of an area in depth.

3.3. Research Design and Strategy

This research is more of descriptive approach with case study design of research. Reliable data were gathered up on how the '*Mezard*' indigenous conflict resolution method is employed. As descriptive research is used to obtain information related to the issue at hand (Kothari, 2004). The target population of this study is mainly from the household elders & farmers and the court officials. This is because the researcher believed that the data needed for this study is fulfilled through an in-depth interview with the above subjects. These respondents were determined based on the data entry of the woreda administration and their repetition in such practices of eldership is considered. In addition youth and women are subjects of this study as a supplement to the key informants. The sample size in this study was 24. Twenty one of them are the household elders and 3 are the court officials, one each from the selen-weha, waja, and timuga *tabias*. The size is determined according to the trend in which similar researches were conducted. For instance, Tsega Endalew, Dejene Gemechu, Melkamu Abate, Zewde Jotte, and the like take sample size of 44, 10, 26, and 33 respectively as key informants. And since in qualitative research the maximum number of interviewee is 20 the researcher was used the proposed sample size. The *tabias* were purposively selected as they are predominantly exercises the *Mezard* institution of conflict resolution.

3.4. Sampling technique and procedure

Three *tabias* i.e. Timuga, waja, and selen-weha were purposively selected. The researcher used non-probability sampling technique and especially the convenience sampling as the researcher selects the household and judge respondents for the convenience of the study and in fact the judgmental sampling as the researcher believes that the selected samples were typical representatives of the whole population. The total population of the woreda is 20,597 households. From this total population 24 respondents were selected as a sample size. From this 21 were elders, 7 from each *tabia*. The remaining three were court officials, one each from the three *tabias*. The interview schedules are developed in English and later translated in to local language Amharic and Tigrigna. In order to collect the primary data from the elders and court officers two individuals helped the researcher in getting the elders to be interviewed since the researcher is not most familiar with the area. Hence, the total sample size is given as;

i. 7 Elders from each *tabia*= a total of 21
ii. 3 court officials, one each from the respective *tabias* = a total of three
iii. Three focus group discussions with ten members each from selen-weha, Timuga, and waja. They constitute individuals who experience some sort of conflict, women and youth.

3.5. Data Collection Instruments

3.5.1. In-depth interview with Key informants

Schedule was developed by the researcher. The interview schedule was developed only after a thorough review of documents, a personal observation and discussion with experts. In order to accomplish the data collection the researcher takes about a month. The elders/opinion leaders, court officials are interviewed using different interview schedules for collecting primary data. In order to collect the primary data from the research site, the researcher conducted an in-depth interview with prominent elders from the *tabias* with an average of two years experience being decision makers in the *Mezard* indigenous conflict resolution institution. Indepth interview was also conducted with court officials selected from the respected tabias. Since the researcher is not most familiar with the area, two individuals from the *waja* and *Timuga tabias* assisted the researcher especially in cooperation in meeting and convincing the resourceful informants.

3.5.2. Focus Group Discussion

In addition to conducting individual interviews with elders, focus group discussions were conducted with the members of the community by considering gender composition to have more information about the cause for conflict with respect to access to justice as well as on how the whole conflict resolution mechanism process takes place by the *Mezard.* There are three groups with ten members each from the three selected *tabias*. Of these groups the fist group was women drown from local community, the second was composed of youth from local community, and the third one is the individuals who were got justice through Mezard. And the discussion was held one time each. Some of the community members, who have accessibility to one or more cases of conflict handled by the institution, youths, and females, are included in the focus group discussion.

3.5.3. Observation

In addition to the above mentioned instruments non participatory observation was conducted on the area by taking notes based on certain checklists. The researcher observed what and how conflicts are resolved practically by the *Mezard* conflict resolution method on the selected *tabias* by conducting non-participatory observations on some cases of conflict resolution. The observation is in an open or visible manner. The researcher also observed the people's style of communication and way of interaction. This was conducted one time each in the selected sites. This method was used in different ways. Through the preparation of check list which enables to obtain information of what an area looks like, attending some of the methods of reconciliation, ways in which the elders handle and process the several types of conflicts.

3.5.4. Document Source

Documents pertaining to the topic of the study were reviewed and analyzed. In fact, documents include many ranges of materials such as books, news paper, magazine, notice, letter, minutes of meetings, diaries and so on and so forth (Robson, 2002). The same author also states that documents could additionally include non-written forms as films, photographs, television programmes, drawings, and pictures. However the researcher uses only the document of *Mezard.* Because of the inconveniencies created photographs and films were not employed.

Documents have their own advantages over the other techniques of data gatherings in such a manner that it is an indirect data collecting tool. The document is not primarily prepared for the purpose of the study as in the questionnaire, observation and interview in which there is a chance for the informants to change their behavior and alter the information in one way or the other. Creswell, (2003), in this regard, also adds that documents:

> • *Enables the researcher to obtain the language and words of the participants*
> • *Can be accessed at a time convenient to the researcher – an unobtrusive source of information*
> • *As written evidence, it saves a researcher the time and expense of transcribing. (Creswell, 2003:187).*

The documents that are related to the *Mezard* and any data related to the communities are apparently serves to acquire information in relation to the issue. Therefore, the researcher critically examines documents that are reserved in the *woredas'* as well as *the tabias* archives for insight evidence on the prevailing problems for the proposal. Secondary data is also used from several literatures and court legislations of the *Mezard* institution by itself since it has a written document. Books, journals and articles are reviewed for obtaining different ideas on the study problems.

3.6. Data Analysis and Interpretation

Data that are collected through observation, interview, and FDG were analyzed qualitatively. The qualitative data is analyzed by making use of descriptions and interpretations. Hence the interviews, discussions, and observations are translated into English and an attempt is made to keep to the original version. The data are presented in a narrative form so as to keep the originality of the research findings.

CHAPTER FOUR: RESULTS AND DISCUSSIONS

Mezard Institution of Conflict Resolution in Rural Alamata Woreda

This chapter briefly identifies the *Mezard* indigenous conflict resolution in rural *Alamata* with focus on *Mezard* institution. It mainly emphasizes the source of conflicts, the institutional development of *Mezard* the mechanism of indigenous conflict resolution, conflict preventive institutions, the jurisdiction of the institution, the process of addressing murder case in *Mezard, enforceability, sustainability* of the institution, the relationship between *Mezard* and formal court as well as strengths and weaknesses of the institution.

4.1. Sources and Type of Conflicts in Rural Alamata woreda

As the researcher has observed the rural *Alamata* woreda is an area which has a population which is engaged predominantly in agricultural activity as most of Ethiopian population with some activity of irrigation together with cattle breeding. This agro-pastoralist population has a strong coherence among themselves, because they are intertwined by several socio-cultural activities and practices which scale up their unity and coherence. Above all they give due emphasis and respect to their indigenous institutions or *Wonzs* which contribute for the peaceful resolution of conflicts.

Conflicts may exist between individuals, families or between groups stemming from misunderstanding or verbal abuse to physical violence including death of parties to the conflict (Francis, 2006). These conflicts can be managed or resolved through various systems, for example, through dialogue or discussion. Hence, various indigenous practices or resolution methods provide in different laws, regulations and rules of institutions, are helpful for settling conflicts.

According to the data gained through interview from the household elders in the study area, there are so many sources of conflicts that emanate from various perspectives. For instance, individuals conflict may stem from his/ her motivation to power like the need to be village head man, between group from need of resources like the rights to water or to pasture or boundaries, possession land, demand for resources (like water, grazing land, etc) which means scarcity of resources are usually reasons of conflicts. Thus these varieties of sources give rise to a conflict to happen, and contributes to socio-economic problems in the area.

All members of the focus group discussion conducted in Waja outlined that there are several sources of conflict in the area. Among them are conflicts raised because of economic interest or scarcity of resources, demand for fairness, and need for social status and the like. These in turn contribute to conflict to break out within the society. Accordingly the members of the focus group discussion outline that social conflicts such as rape, abduction, and land conflicts are dominant in the area. Therefore this shows that socio-economic conflicts take the first hand in the study area. In line with this the researcher observed that the *Mezard* institution is giving services only two times a month. This is however raised by the members of the focus group discussion as it is not worth enough.

There are several types of conflict manifested, observed and reported in the study area. Accordingly, the court official from Waja put it as follows;

> Resource related conflicts such as on ownership of plot of land, debt default, utilization of forest resources, over grazing land, boundaries of farm plots are common in the study area. There are also socio-cultural and value related conflicts such as marriage conflict, sexual offence, insulting an individual or a group of people, extra marital sexual relationship, etc. in addition there are also cases of murder, and conflicts which causes bodily injuries in the study area. But the cases of homicide are extremely rare.

The data obtained through interview conducted with household elder from Timuga signifies that in the institution one of the serious crimes beside the homicide is having a relationship with married women. It was literally described as "*yabawera mist yeneka aras nebir endiyameta ytazezal*", literary means those who made a sexual relationship with married women are obliged to bring a furious tiger. This implies the seriousness of the above practice which totally impossible for the suspect to meet the penalty.

According to the members of the focus group discussion conducted in selen-weha the above types of conflict are really happening in our area. But two of the members of the focus group discussion have a reservation that the *Mezard* institution solves land conflicts especially on the border with Amhara regional state. In their answer to the question if the *Mezard* institution solve these conflicts;

> Of course the Mezard settles all the types of conflicts which are listed above. In case it has specific rules and regulations to guide every conflict happened either cultural, social, value, etc. what the Mezard is not confronting with is the conflicts which are related with politics and political matters. With regard to politics the Mezard institution has a non-interventionist policy.Its practices are out of political issues.

The above issue shows that the institution solves almost all sorts of conflict happened in the locality with the exception of conflicts which are expected to be more sensitive and latter develops in to a political issues. These include religious interventions, highly controversial land issues and the like. Therefore the *Mezard* is a non interventionist institution. It only deals mainly with civil cases.

4.2. Historical inception and Development of Mezard institution

According to the key household informants, *Mezard* is the name of a person and his name is taken as one of the name of *Rayan* "*Wonzs*" which means house of elder fathers. His birth place was a controversial issue as it is discussed by the members of the focus group discussion, some of them trace his birth to *Areza* in Eritrea, while others claim that his birth place is in *Temben*, and then he further came to *Raya* area in specific highland area named as *Kidanemhret* or *Jan Amora.*

During his existence, he disagrees and fought with lowland part of *Raya* Oromo. Initially, *Mezard* could not defeat the lowland Oromos, but later one person called *Abeto* with his soldiers came from northern Tigray to the area and settled in the highland area named as "*Tebaba Ber*". And then, both *Mezard* and *Abeto* tried to discuss each other for collaboration to defeat the low land Oromos. After the victory, *Mezard,* with his neighboring governors *Geberemedhin* and *Boza* spreads indigenous laws in *Imbabrko* to the east of *Waja* specifically in the place known as *Hudad.* These laws are related to criminal laws (e.g. Murder case) and civil laws (e.g. Marriage case). According to oral tradition, the document of their agreement is found in *Kurfa Arfa* a place found to the south of *Waja.*

Mezard is the first conflict resolution in *Raya Alamata* woreda which has a written document. It has also unique character in the sense that it almost does the reconciliation process similarly all across its area of coverage unlike the others. It had also a point of difference on the way of compensation it follows. Since *Mezard* follows a compensation based on the consensus i.e.in accordance with the written document, where as the other institutions follow a compensation based on the amount requested by the family of the deceased.

Mezard was the founding father in areas of arbitration to resolve conflicts and built peaceful coexistence among the people of the area. He had five sons, such as Shum *Zemariam*, *AmdeMichael*, *Ayno*, *Sbho* and *Shahek.* He divided the area and shared power to his sons to settle conflicts being as arbitrators with oral indigenous guiding principles. For example, "*yedem waga*" (blood fee) is equal to the price of one mule or literally "*yeand boklo waga*" is forty birr as compensation for the deceased person.

According to the court official from Waja, since 2007, this institution has transmitted its oral law into clearly identified codified laws. The process of documentation of this indigenous practice is democratic and participatory. All the members of the community participate either directly or indirectly. The member of elder fathers identified harmful traditional practices to be eliminated through education and financial punishments with the support of *Raya* Development Association aiming to improve and strengthen previously existing indigenous practices. According to the document there are newly added ideas in to the document of *Mezard.* Finally, the document was forwarded to the people for further approval and awareness creation about the newly added rules.

This implies that the *Mezard* institution is developed in to being compiled as well as chartered institution since it provides a written material up on the cases handled and the decisions made or passed by the elders. For instance, providing receipt in return to the financial punishments, a letter of decision for the litigants, etc. This can have a good implication in preserving the rules and procedures of *Mezard.* Thereby its sustainability can be maintained.

According to the court official from Timuga suggests that in regard to recognition all the decisions passed by the indigenous tribunals are recognized by the court as long as the conflicting parties are willing to be litigated by the institution. This implies that there is a full recognition of the institution by the court but, not as last and binding.

The *Mezard* indigenous conflict resolution institution is headed by elder fathers each locality in order to resolve individual or group conflicts. Among which *Mezard* is one of the indigenous institutions in Raya that address conflicts of any type. This indigenous institution has local laws promulgated by previous elder fathers of the localities. Currently, these indigenous laws have improved and include many elements. Its codification is carried out under the leading position of *Raya* Development Association, prominent elders, religious leaders both the Muslim and Christian and elder women have participated in the initiation of the codification process. The document of *Mezard is* divided in the two major parts. In the first part of the document there are list of harmful traditional practices which deserve to avoid through lesson. For instance, genital mutilation, tooth extraction, extraction of tonsil and others. In the second part of the document, there are also list of criminal acts and bad practices related to civil acts.

For instance, homicide, rape, abduction, marriage, *teskar* (commemoration for dead) and others. Punishment is a must for the commission of these offences. However, the interview with the household elders affirmed that such practices still exist among the community although set of penalties arranged. This shows that the procedure written on the document of *Mezard* under harmful traditional practices is not implemented as such.

4.2.1. The jurisdiction of the Mezard

The *Mezard* indigenous tribunals have their own judicial or local jurisdiction. *Wonzs* are demarcations for different indigenous institutions involved in resolving conflicts for the given society of the area.

According to the key informants response *Mezard* covers the areas from *selen wuha* in the east up to *jen Amora (kidanemhret*) in the west and from *Gobu river* (in *Waja*) in the south up to *Aroresha wenz* in the north. This institution manages conflicts that may arise in the localities like *Babo korma koba* (include *Botamariam*, *Arkisa*, *Chelekot*, *kubikerensa*, *berbero*, *Adawly*, *Hudad* upto *kidanemhert*), *Waja*, *Timuga, selen-wuha* etc. indigenous tribunals in this institution see civil and criminal matters which are disclosed in the document of Raya Development Association. For example, in article 1 and 2 are about civil matters. Article one includes several harmful indigenous practices such as genital mutilation, early marriage, cutting forest, hunting wild animals, etc. article two is all about the extra resources spent for several ceremonies such as baptism, '*teskar*', marriage and in article 3, 4 and 5 are about criminal matters like rape, theft and homicide cases respectively. Article six and seven is about committing wounds and arranging reconciliation respectively.

According to the document of *Mezard,* in article four of the powers and responsibilities of the indigenous tribunals if the arbitration is failed, the elders of the locality lead the case to the respective indigenous tribunals which are the *Ketena* indigenous elders . In this way the *ketena* elder fathers take into consideration appellate jurisdiction. If the case is not addressed in the both levels, it will be transferred to woreda elder fathers for final decision.

4.2.2. Actors in Mezard institution

According to the interview conducted with the household elders from in the whole process of conflict resolution there are typical actors such as the victim, offender, elders, witnesses, relatives and elders. From these actors elders are the most respected and decisive actors. They are predominantly engaged in reconciliation process. Being an elder in the community is believed as a source of moral satisfaction as being able to solve disputes peacefully. During the process of resolving conflicts the parties have an ample opportunity to their side of the case and to discuss the proposal for final decision. However, age is not the main criteria but ability to negotiate and recognition by the members of the community in being member of the eldership.

4.2.3. The role and power of the elders in Mezard institution

Elders in *Alamata* woreda are respectful for their tremendous function and balanced, fair, and critical decisions in the society. They are of such prestigious status that their deeds and decisions are not simple and easy to refuse. Elders play an important role in the society to properly administer and harmonize the people. This is suggested by the elders in Waja. However, the focus group discussants in the respective *tabias* suggests that that,

> To some extent there are reservations that some members of the indigenous tribunals are biased. This is seen for instance in accordance with the decisions given for women and men is not equal or fair. There is also some sort of favoritisms or inclination for relatives as well as giving priority to prestigious individuals. Hence some sort of accepting bribes is also seen among some members of the indigenous tribunals.

Members of the FGD also disclosed that, the request made by the elders for the offender to bring and prepare food and *tela* or *tej* and other facilities for the elders is also seen one of the short coming of the elders. Because, there are individuals who are incapable to meet the demand of the elders as a result of several socio-economic problems. Therefore this idea contravenes with the above stated characteristics of the elders.

According to the interview conducted with one of the most prominent elders in Waja, the role of the elders is to carefully argue for certain issues and convince the conflicting parties to accept the proposed alternative of decision. This means that, the major way is persuasion rather than threat of force. Hence the elders look for win-win solution and finally they are aiming at repairing severed relations among the disputants based on truth, conviction or remorse, sanction, accountability for justice and forgiveness. In contrast with this, from the members of focus group discussion especially three youths and one woman argue that, there are cases where the decisions given by the elders are not in a position to be win-win rather win-lose or lose-lose. For instance they refer some cases as the victim is advised not to ask compensation simply for the sake of guarantying reconciliation. This is most frequently seen in the conflicts over natural recourses such as grass, cattle, or something else. Thus this implies that the decisions given by the elders sometimes create dissatisfaction among the members of the community.

Every matters of the society such as marriage, funerals, baptism, etc are managed by the elders who are locally called '*shimaglewoch*' or '*yewonz abatoch*'. This can be synonymously being referred to as the council of elders. Members of this council of elders are elected in a seemingly free, but partial democracy by their respective community. As it is discussed by all the members of the focus group discussion, it is partial because women and youngsters are not allowed to participate in the process of election both to elect and to be elected. In deed women are given delegations and such a full- fledged power to settle every kind of problems that might arise in the community. They further argued that the council of elders does not have an absolute decision making power over the community. Hence, they are primarily accountable to their public which elected them. If in such cases the public at large loses trust on any member of the council of elders, it would have then the right of recalling.

As to the elder interviewees, indigenous tribunals don't have the authority to mediate the disputants to bring about reconciliation. Instead, they facilitate the situation to get calmed through the involvement of the third party. It is therefore up to the will of the disputants to elect mediators. They believe that the elders who are elected to get involved in the process of mediation need to fulfill the age requirement of 50 years and above in addition to the high respect and trust among the community to mediate cases. While the rest argued that the age doesn't matter as long as one has the required knowledge. But all agreed that the elders should not be closest relatives of the disputants. This is done basically because of the conviction that the process of mediation and the final result of the verdict ought to be free of bias and any tendency of nepotism.

The interview conducted with the household elders shows that elders are chosen for membership of the council for their high moral standards in the community. Being relatively aged in the community is also another criterion for being a member of the council. This is understood by the society as their knowledge on particular customs, skill and wisdom in handling disputes and an ability to analyze and advise disputants being patient, impartial, free from favoritism, bribery, and respectful to diversity. This means that, from the point of view of the respondents the elders must incorporate characteristics such as fear of God, well informed, tolerance, wisdom, courage, being humble, loving, civil, patient, truthful, etc. however the focus group discussants in the Waja suggests that in practice there are some irregularities created as a result of biases and favoritism made by some members of the indigenous tribunals towards the rich.

As Zartman 2000, contends the most important elements involving in gacaca indigenous conflict resolution mechanism in Rwanda include the tradition of forgiveness, respect for elders because of their symbolic authority to enforce decisions and transfer of resource as compensation. Therefore my study is consistent with the above argument as the *Mezard* institution possesses the character of respect, trust, consensus, etc both by the members of the community and the elders.

4.3. Procedures of the Mezard institution

The *Rayans* have their own indigenous conflict resolution institutions. Those institutions are their own great role in resolving conflicts in peaceful manner. Among which *Mezard* is one of the indigenous institutions in *Raya* that addresses conflict of any nature. This indigenous institution has local laws. Currently, these indigenous laws have improved and include many elements. Its codification is carried out under the leading position of *Raya* Development Association, prominent elders, religious leaders both the Christian and Muslim and elders of women (the *Duberti*) are participated in the codification process of the *Mezard* institution.

The document of *Mezard* is categorized in to two major parts. In the first part of the document, there are listed harmful indigenous practices which are proclaimed to avoid through lesson, for instance, polluting the river by throwing rubbish things to the river, genital mutilation, tooth extraction, being idle and expectance, chewing *chat*, extraction of tonsil and others. This are expected to be solved only through continuous awareness creation and education. In the second part of the document, there are lists of criminal acts and bad traditional practices related to civil acts, for instance, homicide, rape, abduction, marriage, threatening individuals using several materials, etc. these practices are accordingly followed by a punishment or else warning and consult as to the type of the practices. This shows that the *Mezard* has a framework of articles and procedures which contributes for protecting conflicts from being happening as well as providing measures to the wrong doings.

Moreover, the number of elders involved in the process of reconciling the conflicting parties is always five. This is mainly because of the historical inception of the *Mezard* institution, since the person called *Mezard* had five children.

According to the interview conducted with household elder from Timuga, the members of indigenous tribunals of this institution were elected by the people of the locality. Now the ability of a person to solve complicated problems is becoming the criteria for election. Article 2 and 3 of the document of *Mezard* stated that "*Wonzs*" or indigenous tribunals are elected democratically by the people in three tires.

In the village level three elder fathers are elected, in *ketena* level also six elder fathers are elected and in the Woreda level five elder fathers are elected. However the researcher was relied on the elders who are elected at the *tabia* level. In article 4 of the document, the power and responsibility of the three stages has stated. In addition to this, in article 5 and 6 the term of mandate and period of election are stated respectively in the document of *Mezard.* Consistently article 5.1 refers that these indigenous tribunals shall be elected respectively if they own majority vote. Article 6.1, 6.2 and 6.3 are concerning the electoral period which is conducted in the village, ketene, and woreda levels. Accordingly, two, three and four are terms of election respectively. In this document there other organs of the institution are put with their function and responsibility. This includes the audit committee and the *Raya* development association.

Regarding the selection criteria, one of the household interviewee has explicitly stated as follow:

> Being mature enough in age wise is not a sole prerequisite to be an arbitrator. Being only hero itself is not led to full respect.Having a wise talent to tackle and solve complicated socio-economic problems among the society and having all-rounded knowledge is required to be an arbitrator.

Whereas the document stresses that the elders at the *tabia* level are elected directly by the people and they in turn elect representatives to *ketena* as well as woreda levels. This is outlined in article 3 of the power and responsibilities of the elders in the document.

In *Mezard,* the procedure of conflict resolution is done as the family of the offender or the offender himself or herself primarily send religious leaders either can be a priest in Orthodox Christian or Sheik in Islam or else elder of the locality to beg for the "*erq*" (resolution). If the victim or victims family are interested and willing, the conflict among the offender and the victim will be resolved. Hence, the elders or religious leaders arrange conditions to make the compensation payable. But if the arbitration fails, the elders of the locality lead the case to the respective indigenous tribunals which are to *ketena* indigenous tribunals. In this way the Zone elder fathers take in to consideration the right to appeal. If the case is not addressed in both levels, it will be transferred to woreda (district) elder fathers for final decision in the *Mezard.* The *Mezard* institution does not tolerate for the person who committed any wrong action and ignores taking his case before indigenous conflict resolving bodies. In *Mezard,* normally the families of the offender or the offender him/her self primarily send religious leaders they can be in orthodox Christian (priest) or in Muslim (sheik) or elders of the locality to begin the "*erq*" (resolution). If the victim or victim's families are interested, conflict will be resolved and then the elders or religious leaders arrange conditions to make the compensation payable.

In relation to the crimes which are conducted or committed by underage individuals, one of my interviewee contends that, in such cases the child is sent free, but the compensation or decision will be binding up on his/her families.

4.3.1. Type of Murder Case in Mezard

In addition to the above procedures indigenous tribunals of the locality, see murder case by categorizing into two ways as "*key dem*" and "*tikur dem*". This is equally put in the document of *Mezard.*

'*Key dem'* (red blood) occurs when the commission of murder is made negligently or accidentally whereas "*Tikur dem*" (black blood) shows situation by the murder has committed the offense of homicide allegedly and intentionally. Such categorization is not merely for theoretical purpose but also crucial for determining the penalty to be imposed up on the wrong doer as well as the amount of compensation to be paid to one victim family. As a result, it is used to mitigate the situation of the punishment. In addition to the above categories one household interviewee stated that there is a condition in a murder case where the slayer commits a murder severely which gives rise to the pre-established compensation fee to become double. The range of compensation to be paid is 15,000 and 20,000 for *key dem* and *tikur dem* respectively. The variety is created because the earlier is committed negligently and unintentionally. For instance, the case of self defense. According to the household respondents the practice and implementation is effective. However there is a concern on the side of the focus group discussants that individuals with some amount of wealth can commit crime again and again. Therefore, this is less significant for the enforceability of the decisions up on the murder case.

4.3.2. The process of addressing in the murder case in Mezard

While a murder or homicide is committed the procedures of reconciliation are typically as follows:

According to the interview with house hold elders the one who commits a crime of homicide is expected firstly to admit to the police or security. Then latter his case began to be seen by the elders. He is expected to have an agent to reconcile with the elders on behalf of him. In addition, the slayer's brothers, and his close relatives are expected to leave their home village temporarily until the reconciliation process is finalized. This is mainly for the purpose of protecting the possible retaliation.

According to one of my informants, most of the time *erq* is made most commonly after 40 days from the death with relatives and one year with the slayer. Till then, the slayer needs to disappear from the locality whether he/she committed the crime intentionally or by negligence. In addition to this, indigenous tribunals of the locality, see homicide case by categorizing in to two as *key dem* (*red blood*) or *tikur dem* (*black blood*).

The above practice implies that there is less or weak approach to control the slayer or the offender. Because there is no any guarantee for the offender to be around the area until the *erq* process is conducted. This can result in the disappearance of the slayer elsewhere without fulfilling reconciliation, so that there is a need for the court to intervene in such cases.

In regard to the reconciliation process, the family and relatives of the offender appeal to *Mezard* and religious fathers. Then the offender's families send religious leaders and elder fathers of *Mezard* to the victim's family.

As one of my interviewee suggest, in the case of the Muslim religion, the Muslim religious leaders plant a tent around the place where the deceased families are dwelling. They drummed a drum, in a sense to mean that, forgive us the offender or his or her families. According to oral tradition in Raya, Sheiks have spiritual power which is given from their fore father to bless and curse even achieving it. In this case, the neighboring people would enforce the victim family to be volunteer by saying do not blame the deceased by elder fathers and religious leaders. For the Christians especially the orthodox, during the time of resolution, according to the informants, religious leaders, elders and women have a great role. For example, the religious fathers by handling the cross and church umbrella try to beg and convince for the deceased family with chanting by means of "*Igziomeharene*" (God forgive us). Usually the "*erq*" process is made around a church and under a big tree mostly *wanza*. These practices indicate forgiveness for the slayer and his families. If the victim family are interested conflict will be resolved, then before the final decision of the elders an oath is taken by religious leaders and elder fathers.

In line with this, a group of women are known as "*Duberti*" hold a red stick cutting from small stick of "*wanza*" tree they conduct a kind of begging and requesting the so called "*mereba*" to bring the conflicting bodies especially the slayer and the family of the deceased to come in to negotiation and reconciliation. Thus women have a significant role in bringing the conflicting parties in a round chair. The role of *Duberty* (group of women) in the *Demerq* (blood resolution) process on the other hand is very significant. The elder women are very much respected and they have a significant place in the community. Women would hold blood resolution together with elder fathers and religious leaders. They hold a red stick cutting from a small branch of "*Wanza*" and they make a kind of traditional request which is called "*mereba*." The implication is begging the family of the deceased to forgive the slayer or his family.

As it is discussed by the household interviewees during the time of reconciliation, the murderer is brought before the family of the deceased, wearing a thick black cloth locally called '*mak*' accompanied by neutral third party groups of some '*Militias*' in the presence of religious leaders and elder fathers. The piece of black cloth implies humiliation for what the person has done. It is one of the mechanisms to punish individuals for their wrong doing. In fact, it is believed that it is also a mechanism by which the criminal is allowed to appeal for forgiveness by wearing the thick black cloth the *mak*. For this purpose this practice made the criminal as weak and helpless among the locality. This perception implies still the attitude of the community over females is bad, since they consider women as inferior and weak to men. Therefore this can be one of the implications for the women not to participate directly on the decision making process in the institution. The criminal is then requested to lie on the ground and the family of the deceased and asked if they want him killed right away. But the family of deceased actually answers the opposite. The response by the victim's family is, as follows "we do have children who might in the future do the same by mistake, therefore, and we do not wish a death sentence to our children in case that happens". This shows that the members of the community are highly respected to the decisions given by the elders, as they believe they are the source of peace and integrity.

Following this, all family and relatives of deceased person in the '*dem erq'* (blood resolution) would stand up together and say, one after the other "we forgive you for the sake of God". *Wonz* and religious leaders would then stand and bless the family by saying. "it is good for us to learn that, actually you are God's people. Therefore, God bless the remaining of your family and may God give you peace and love with the wrong doer.

One household elder interviewee also disclosed that, before the end of the practice of *Wonz* proceedings, all relatives of the deceased are required by the council of elders to remain under an oath not to attempt retaliation in order to avoid killings in the sense of revenge. The oath is taken by touching the true cross of Christians or the Drums of Muslims. Sometimes this Oath is performed by touching the ground.
He also added that a murder case is not in fact the most frequently seen practice in the area as in the other social conflicts. As to him from the reconciliation process on different cases conducted and handled with an average of two times per month the murder case takes place in a very few interval. Accordingly the rate of interval is ones or sometimes twice per annum.

This implies that the elders in both Christians and Muslims are highly respected. But, the families of the deceased have here a significant contribution for the reconciliation process to be fruitful and successful by due respect of the elders in place.

After these all processes, the victim family will give up to reconciliation and take a promise by the words of religious leaders and elder father to arrange time and place for carrying out the "*erq*" or resolution. This implies that the community's respect for religion as well as the elder fathers brings a safe way for the willingness of the victim's families to reconcile and further forgive the slayer and his families.

Finally, as the initial preliminary procedure is conducted including oath taking then the offender's family and the victim's family eat '*Anebabro*' or *injera* together in one place to have. This helps them to have the spirit of unity and oneness, to avoid future conflict and for the purpose of kinship, etc.

4.3.3. The Compensation System in Mezard

Regarding compensation, the slayer is for the time being expected to pay some amount of money. The rest is reserved for the remaining compensation and the offender brings a man or women as a guarantee called "*Meyazna*", then the date will be fixed and payment effected accordingly. This method differ the institution from the others in the sense that the others request full payment of the compensation at a time.

The indigenous institution of *Mezard* has provided varied amount of compensation for each damages. For instance, article 5 of *Mezard* stated 15,000 birr for "*key dem*" whereas 20,000 birr for "*Tikur dem*". If the same person committed a homicide for the second time, the murderer is required to stay ten years even to organize the '*erq*' process and to pay 40,000 birr. This shows that there is a severe punishment in return for the individual who commit crime again and again, and it warns individuals to be cautious for their practices. According to article 7 of *Mezard* the '*erq*' process of '*key dem'* is held in the time of funeral. After all, to make the '*erq*' very stronger and binding the parties will create kinship through inter-marriage and baptism. This situation is facilitated by the middle party or other elder fathers.

In some cases after compensation is paid, in the case of "*Tikur dem*" there is a practice where the murderer is made to live in place far away from the locality of the victim's family. This tradition is expressed in article 7, sub article 6 of *Mezard.* Whereas if the murder is found to be "*key dem*", the disputant may establish a smooth relationship with each other through marriage, baptism and other value bonds in order to make them feel sense of forgiveness.

The other way of reconciliation as discussed by one of my interviewee is the so called "*yerejat*" or the reconciliation of the near family. Here there are two mechanisms of reconciliation, either in money or physical punishments. According to the focus group discussion comprising of the individuals who have an exposure in the *Mezard* institution either as offenders or victims there is a specified compensation of money for all near relatives of the slayer to pay for the family of the deceased. Whereas, the mother and father are to be reconciled without any payment. The other method is through physical punishment. Hence, there are individuals whom they have to be beaten by stick, stone, etc so as to accomplish the process of reconciliation. In this case those who are not near relatives of the slayer are categorized in to; those in which stone is thrown to them, those who are beaten by stick, etc. by the relatives of the deceased in order to accomplish the *erq* process. This tradition is somehow not fair as it violates the human rights of individuals.

The people of *Raya Alamata woreda* have therefore a tradition to cooperate one another in time of hardship, mostly when the slayer or his relatives are unable to pay the amount of compensation so decided the slayer is obliged to beg money in the place where many people are available like in the market, church and mosque. The people in the area in turn contribute money to the person.

4.4. Enforceability of the Mezard institution of conflict resolution

Concerning the enforceability of the *Mezard* institution the members of the community have a consensus to be respected and abided accordingly by the laws and procedures of the *Wonz* of *Mezard.* Its full enforceability is based on the value and status given to the elders by members of the community. Otherwise whenever the decision of the elders is violated it is hard for the offender to live harmoniously among the community. This is because he/she is out casted by both the elders as well as the members of the community from any kind of social relationship. On the other hand the court also has the practice of accepting the decisions passed by the elders. In such cases it recommends the conflicting parties to be abided by the decisions made by the institution. As such it has no role in implementing the decisions of the elders.

The participants as well as the conflicting parties who come to the *Mezard* institution are as a result convinced on the basis of various sources and witnesses to be abided by the final decision. Whenever there is a party which fails to be abided by the final decision a serious sanction is being followed by the members of the community. These are curse, prohibiting from any social festivals, discrimination from the community and other psychologically painful punishments unlike the formal court having financially and physically painful punishments. Accordingly one of the household elder informants said that;

> Since the custom of the institution i.e. wonzs and the elders are highly respected by the members of the community most of the decisions made are easily implemented.

The members of the community mostly prefer to refer the *yewonz abatoch* whenever they face conflicts. They need adjudication and latter reconciliation through the *Mezard* institution. One of my interviewee in his words confirmed that;

> This arises from its high acceptance and legitimacy from the people as it shares the following characteristics such as cost effectiveness, accessibility and proximity, emphasizing on reconciliation, familiarity, wide participation where decisions are made through arguments and discussion by all concerned bodies.

The indigenous conflict resolution mechanism in *Raya Alamata* had its own ways for implementing decision. As it has been mentioned in the preceding section, after the practice of "*erq*" is concluded the offender is obliged to pay the amount of compensation as determined by the indigenous tribunals within a specific period of time.

Comparing the cases which are seen and handled by the modern court with the *Mezard* institution one of the *Tabia* court officials suggests that the majority of the cases are adjudicated by the *Mezard* through the indigenous tribunals although numerical explanations are not done by until now. In line with this the court also accepts the decisions which are passed by the elders of the institution. This shows that the court gives priority to the indigenous tribunals as it call the conflicting parties if they come by the *Mezard* institution.

The decision of indigenous conflict resolving bodies is legally and morally binding. The fact that indigenous tribunals would hold decision by ordering the offender to bring guarantee in case when the offender may not apply the decision. But due to social condemnation and exclusion, the decisions of the elder are always executed. This shows that the respect of the members of the community to the elders there by the institution gives rise to prosecution and protects the possible retaliation by relatives of the victim. In the focus group discussion respondents describes this as;

> The enforceability of the decisions made by the Mezard institution seems to be high. This is because people irrespective of their varying social backgrounds consider eldership, its principles, procedures, and outcomes as lawful and binding. So the people largely recognize the value of elders and their reconciliatory role.

Therefore, after a person has agreed to address his case before indigenous tribunals, it is a must for the person to comply with the decision passed by elders even if he/she is dissatisfied with such decisions.

In line with this one of my interviewee disclosed that the right of appeal is guaranteed for the parties who are dissatisfied. But the right of appeal is mostly to the other stages of *Mezard* such as the zone or woreda, and it is based on the consent of the individual. Therefore going back to the formal court after a decision is passed by the elders is not seen and the case is accomplished within the three stages of the institution. As a result, these indigenous conflict resolution mechanisms contribute a lot for the peaceful coexistence of the *Raya* people and their neighborhood.

According to Paul (1994), implementing this approach does not require sophisticated party structures or expensive campaigns; it provides a low-cost, empowering means of resolving conflicts within a relatively short time frame. In many societies, elders have indigenous jurisdiction in facilitation, arbitration, and monitoring outcomes. This is similar with my study in the sense that the *Mezard* institution saves cost, time and energy of the litigants which makes it preferable by the members of the community.

As Hamer (1987), disclosed that indigenous institutions of conflict resolution impose sanctions which are of two types, the secular and the sacred from which the later is more effective and often used. My thesis is consistent with the above issue except on the effectiveness of the sacred sanctions. This can be seen for instance, in the case of reconciliation the institution strives to convince the victim to come in to arbitration in the name of God, but in the case of secular sanctions the *Mezard* institution has a success story especially with the psychologically painful sanctions much as the sacred sanctions did.

Pospisil (1990) proposes, if the sanctions made by the indigenous institutions really more important and binding, appealing to functional approach that what actually qualifies sanction is its effective social protection. This is also true with *Mezard* institution as the decision passed by the elders is binding up on the conflicting parties. The right to appeal is mainly reserved up to the third stage of the *Mezard* ie. the *ketena* level. In case there appeal goes further to the informal court, the court simply decide the parties to be compelled with the decisions passed by the elders. This shows that the formal court didn't make different decision again.

4.5. Relation between Mezard and formal Court

According to the court officials, the relation between indigenous tribunals and state justice authorities is based on mutual cooperation. Moreover, the relation between *Mezard* and state court authorities of *Alamata Woreda* has become positive and strong, because both of them have used *Raya* Development Association as a bridge between them to work closely and to have a sense of cooperation. The raya development association has a significant role in convincing the elders throughout the study area to accept the compiled document of *Mezard* and to perform their activities based on the procedures outlined in the document. This organization is careful to collect and compile the fragmented procedures and decisions made by the *Mezard* previously in order not to lose their indigenous character. And of course all the elders across the study area, women and religious leaders discuss on it and finally accept it and currently it is fully functional in the study area.

Raya Development Association currently works to avoid harmful traditional practices in *Alamata woreda* by using indigenous institutions like *Mezard* as a tool to achieve its objectives. As a result it is better for the people of *woreda* to support these institutions as well as *Raya* Development Association for the well-being and peaceful coexistence of their own.

The written law of the *Mezard* goes beyond the individual rights far to group rights. Accordingly it suits itself with the socio-cultural and economic aspects of members of the community. According to the document which has compiled regulations and procedures to guide the institution there are specific regulations with regard to marriage, baptism, harmful cultural practices, physical damage, murder, insult, etc.

With regard to the question if there is any support given to the institution by the court, one of the court officials disclosed that until now there is no any significant support given to the elders beside organizing them together with other bodies such as youth, religious leaders, etc specifically to work on reconciliation in the urban *Alamata* woreda. This shows that there is weak support in facilities from the side of the government to help the elders to enhance and develop their capacity of arbitration and reconciliation.

The laws of the *Mezard* institution of conflict resolution are different in its scope from the formal or modern court. The FDRE constitution under article 34(5) and 78(5) recognizes the application of indigenous laws in civil and personal matters, provided that the conflicting parties agree to the jurisdiction of indigenous tribunals. In spite of the above idea, one of the elder interviewee suggested that;

> Mezard institution exists in a full capacity and handles a considerable number of cases such as criminal, civil, and commercial in nature. Accordingly a larger number of cases are settled by the Mezard tribunals than the formal or regular courts.

It is therefore believed that an institutionalization of the *Mezard* conflict resolution mechanism and bring it in coherence and relation with the formal court is needed for the proper functioning of justice.

4.6. Sustainability of the Mezard institution of conflict resolution

There is a concern that *Mezard* indigenous conflict resolution as it is traced above found to change its structure and indigenousness especially in the urban areas like *Alamata* town due to some forces of modernization. This is mainly because of the introduction of the formal court system which creates some irregularity with the existing indigenous method of conflict resolution to have fewer acceptances with some members of the community especially the young generation. This led to the introduction of new institutions such as the reconciliation committee and community forum. As to the *Waja Tabia* court official;

> These institutions incorporate individuals from different section of the societies such as teachers, women, residents, elders, youths, priests, sheiks, etc. These are organized within the frame work of kebeles, tabias, and blocks within the above structures i.e. the reconciliation committee literally named as Abo selam ,Abahagas and the community forum.

Moreover, the formal court system and the indigenous way of conflict resolution are somehow tending to complement each other. Another court official clearly put this as;

> The court permits the elders to undertake cases out of formal legal structures to settle them using indigenous conflict resolution based on the consent of the conflicting parties and the court. Even when we observe in the Alamata town there are individuals who are engaged in indigenous eldership, settling cases out of the formal court.

However, the sustainability of the *Mezard* is not in a promising condition as one of my interviewee contends that;

> During the process of conflict resolution there are participants such as a group of youth, a group of women, religious leaders but they have no the power of decision making. Although this can help especially the youth to acquire the knowledge on the practice of negotiation and arbitration through the Mezard institution it cannot be a lasting solution to transfer the institution's practices to the next generation as original as they are.

As it is outlined by DFID, the future of indigenous conflict resolution mechanism can be seen as hopefully. Because many states began to adopt them in line with the formal court litigation. Briefings among them include some of the country experiences such as *Shalish in* Bangladesh *,rondas compensinas in* Peru, *chiefs and assistant chiefs in eastern* Kenya*, street commities in* South Africa*, etc.* the above finding is also true in this research work as the *Mezard* institution and other institutions which are of similar future such as the *abahagas*, *Ronds*, *aboselam* etc began to be organized and to some extent institutionalized and play a significant role in peaceful resolution of the conflicts in the area. According to the household interview, in terms of indigenousness the *Mezard* seem to be changed from time to time although the enforcement mechanisms, effectiveness and acceptance by the majority of the community members is in a good manner.

However, according to Lowry (1995), Indigenous conflict resolution's potential effectiveness is diminished where indigenous authority has eroded and the formal authority has increased, since these trends run counter to indigenous values and ways of social organization, including those of handling conflict. This argument is similar with my findings, although there is a concern that *Mezard* institution can change its structure and indigenousness especially in the urban areas because of some forces of modernization has still effective and acceptable among the members of the community. Still its authority is highly respected.

4.7. Strengths and Weaknesses of the Mezard

Indigenous conflict resolution mechanism has their own merit and demerit in resolving conflict by indigenous elder fathers and traditional ways at grass root level.

4.7.1. Strengths of the Institution

According to key informants (elders) and the majority of the members of the focus group discussion argue that the *Mezard* indigenous conflict resolution mechanisms have important contributions. It is an institution which is respected and trusted more than other institutions. This can be seen the community's overall socio-economic activities are tied up with the name of Mezard. For instance they conduct an oath in the name of *Mezard*, their marriage ceremony and issues, baptism and any other festivities are attached with the name of the *Mezard Wonz*. To add some of the strengths of the *Mezard* institution are listed as follows.

Justice is always made in each locality and the judges are elected directly and democratically by the people. In the same manner, indigenous tribunals of *Mezard* are directly elected by the people. As a result of this, comparing to formal court proceedings, this institution is open and speedy trial.

The institution (*Mezard)* system is cost effective way of adjudicating cases. The *Wereda* court is found around 10-15 km far from the locality. But the people would not incur any cost for transportation, legal advisor and other extra costs by using their own indigenous way of addressing conflicts.

The *Mezard* Indigenous institutions support the government in maintaining peace and stability in the locality and its surroundings. This is done by addressing the cases as immediate as possible unlike the formal court.

Moreover, the *Mezard* indigenous conflict resolution mechanisms have important contribution for wrong order. In the process of compensation the victim's family made compromise by reducing some amount of money in the name of all families of the "*erq*" agreement. For instance, for cross, for sheik, for elder fathers and for '*Duberti*'.

The *Mezard* Indigenous conflict resolution mechanism has its own positive outcomes which could improve the economic wellbeing and peaceful coexistence of the community in the area. So members of the community themselves need to cooperate to sustain such spending cultural conflict resolution institution.

Assefa (2005), has summarized only some of the advantages of indigenous conflict resolution mechanisms as they quickly respond to crisis in terms of time, they contribute to the reduction of regular court case loads, they contribute to saving of the public money, it also minimizes the problem in shortage of judges who work in the regular courts, and budget constraints, give access to many people who do not find modern system of conflict resolution comfortable, affordable or suitable to their needs. Accordingly my finding suit the above finding since the *Mezard* institution saves time, money, energy etc.

4.7.2. Weaknesses of the Institution

According to the focus group discussion conducted, these indigenous resolution mechanisms also have their own negative outcomes. Some of them are as follows.

There is no salary or allowance given to them for their day-to-day activity. Even there is no budgetary support from the people due to this the elder fathers are becoming less sensitive. For instance, members of the Indigenous tribunal of *Mezard* are performing their responsibility only two days in a month.

Sometimes indigenous tribunals might be biased to wealthy offenders or victims in exchange for gifts in terms of money and other kinds. Some of the elder fathers dominated by kinship and lack of consensus to pass decision

In the process of "*dem erq*" (blood resolution) the compensation payable to the victim family does not only affect the actual offender but also his relatives. Therefore, all close relatives of the offender would pay some amount of money as determined by the indigenous tribunals in time when locality called "*yerejat*" (nearest family) '*erq*'. Whereas on the legal aspect there is no such predetermined compensation to be paid by the relatives of the offender. The punishment is only targeted towards the offender

In relation to the murder case there is an enforcement mechanism decided by the indigenous tribunals in consultation with the family of deceased to make the murder to leave his original locality far away from the victim family. This can make the "*erq*" agreement to be cancelled, as it lacks safety or guarantee.

According to Lowery (1995), some indigenous conflict resolution efforts may be weakened by age or gender bias for example, in cases with no women elders, some women may believe that male elders are biased against women and that this will be reflected in their decisions. The above finding is congruent with my findings. Prohibition of the women as well as youngsters from direct participation in the institution can bring problems to the *Mezard* institution and further can endanger its sustainability.

CHAPTER FIVE: CONCLUSIONS AND RECOMMENDATIONS

5.1. Conclusions

As it is shown in the in the literature review, scarcity of resources and hence competition to control them causes conflict among people. A conflict exists whenever incompatible interest occurs among individuals, groups and nations, etc. In this sense conflict is an inevitable feature of our social existence. This fact also holds true among the *Raya* people. The *Raya Alamata* people are categorized as an agro-pastoralists people. Thus grazing land is an important resource among the society. The people also practice widely agricultural activities including irrigation methods beside the rain fed agricultural activity.

The *Rayans* have their own indigenous conflict resolution mechanisms that help the society to maintain their solidarity and cohesiveness among themselves. The *Mezard* system serves to control harmful traditional practices and list of criminal acts which are specified with a range of compensations. It is an alternative mechanism to resolve conflict at grass root level based on the interest of the parties without the help of external bodies. It also avoids grievances between the parties in the conflict and maintains their bondage relationship. It is cost effective way of adjudication of cases. In general it is essential for peaceful coexistence of the people in the area.

However, the finding shows that the practicability of the range of compensations arranged for several bad traditional practices on the document of *Mezard* is not significant. This may arise partly from the lack of proper follow-ups by the concerned bodies to implement the penalties levied on the harmful traditional practices. But to the larger extent it is tied up with the awareness and acceptability of the members of the community.

Regardless of the source and seriousness of conflict, the local people have preferred indigenous mechanisms of conflict resolution to solve their problems. All kinds of conflicts except those political matters ranging from insult up to homicide are resolved under the inspection and rule of the *Mezard indigenous* conflict resolution through the intervention of elders. In such cases elders have therefore pivotal and irreplaceable roles in the society. They are the prominent actors to cool down a conflict by managing it from spreading by convincing the conflicting parties to come into negotiation and then to arbitration. The only conflicts which are not solved by the *Mezard* are those conflicts which are expected to be more sensitive to politics.

According to the findings the decisions passed by the *Mezard* indigenous conflict resolution mechanism is recognized by the court. Even at the time of appeal the court is obliged to enforce the decisions which are made by the elders. This means they do not handle the cases which are seen by the elders again from the beginning.

Mezard as an indigenous conflict resolution mechanism institution is effective that it is most of the time a win -win approach, although there are some drawbacks. The aim of decision made by the elders is not basically to punish the criminal nor is to exclude, but integrate him to the society and sustain further peace. It ensures an ease of intermixing the conflicting parties. Any conflict happened between and among individuals in the *Rayan* community is not only left to be the matter of the conflicting parties but the concern of all members of the community, as their mode of living is highly interdependent.

Most of the people in the study area do not prefer to go to the modern court; instead they prefer the *Mezard* indigenous court tribunals to settle their conflicts. They opt to handle their conflict through the intervention of fellow elders at a grass-root level. People choose it for the reason that they believe the modern court fails to reconcile the disputants and their families to end the conflict or enmity once and for all. So, unless conflicting parties are reconciled through the *Mezard* indigenous institution it is most probable and inevitable for conflict to re-emerge which in turn, would disrupt the social relationships and results in disintegration of the society. The modern court unlike the indigenous is also time consuming and financially unaffordable to many individuals. Moreover the judges in the *Mezard* indigenous institution are elected based on the consent of the conflicting parties. Indeed this acquaintance has such a positive role in contributing to increase the reliability of the decision on the part of the disputants and so they are comfortable to accept it. For this reason, the *Mezard* institution has an important place in the elimination of social conflicts in the *raya Alamata woreda.*

Moreover, findings indicate that the *Mezard* system is cost effective way of adjudicating cases. The woreda court is found 15km far from the locality. But the people would not incur any cost for transportation, legal advisor and other extra costs by using their own indigenous way of addressing conflicts. In addition to this indigenous tribunals are looking cases without payment or allowance which might be incurred in the court proceedings. The disputants have the probability to present their case and answer orally. But in the formal courts this particular conflict resolving institution, it is mandatory to present any case in a written form since they are working with codified indigenous rules and regulations.

The sustainability of the *Mezard* institution of conflict resolution can be seen from two angles. First, it is found to change its indigeneity especially in the urban areas because of the expansion of modern courts and forces of modernization. There is also a concern that its sustainability is not worth as the only decision makers are the elders with no proper transformation of the practice to the new generation. Regarding enforceability, the decisions passed by the elders are highly binding up on the members of the society. This is raised from the respect for the elders and also the psychologically painful sanctions by the members of the community. In addition to this the court also justifies and prioritizes the decision given by the elders while cases of appeal are coming.

The *Mezard* institution seems to be characterized as a democratic institution at least apparently. In the real sense however, it lacks an inherently full-fledged democratic element. It in fact appears to encourage participation of the public in general, but excluding the youth and the women to participate throughout the process and in election of the elders who can handle the existing conflicts. Women are only participating in a ceremony called *Duberti* to beg the conflicting parties to come in to negotiation.

The method of conflict resolution by the people on *Alamata woreda* has its own challenge. Indigenous tribunals who are working in this institution might be biased in passing decisions since there is a highly deep rooted value given for the respect of one's own kinship in the area. Sometimes this mechanism becomes a setback to formal legal proceedings by protecting the offender from legal punishment which is decided by the court of low. The punishment passed by indigenous courts mainly takes the form of money, that entails anybody with sufficient wealth can commit crime and live comfortably within the society. It has also significance for the offender by sharing the range of compensation among the close relatives of the offender especially in the case of Murder. In some cases the *Mezard* Indigenous way of addressing conflicts has conducted against the human right of offender and the person in relation to his/her domicile right. Even it has also invisible economical problem for indigenous tribunals themselves.

5.2. Recommendations

If this indigenous conflict resolving institution obtained financial, material, skill and moral support from the people and the government, it will have greater contribution for peaceful coexistence of the people in general and the economic prosperity of the area in particular. Finance can be generated from the community itself and raising the amount of money collected from the offenders. Beside this, generating finance from the members of the Community helps for making indigenous tribunals of the institution to be sensitive and owing due consideration to cases brought before them.

Problems should be maintained through lesson and awareness creation in the community in general. This can be done by establishing two or three forums within each year and making the community aware about the existing benefits of the institution.

Since the Mezard institution of conflict resolution has a well defined procedure, cost-effective practice, and maintaining the social bondage of the members of the community it should be preserved and encouraged both by the members of the community and the administrative agencies.

Since the positive outcome of the *Mezard* institution prevails than its draw backs, this indigenous practice should be maintained. In order to maintain the *Mezard* institution stakeholders should participate in the overall process of conflict resolution through;

- ✓ Knowledge transferring from the elders to the new generation orally during public gatherings like; *equb*, *edir*, *teskar*, etc.

- ✓ Elder's forum for discussing on how to maintain the indigenous institution.

- ✓ The local government should give due emphasis to the indigenous institution by creating some amount of income for the elders who have been serving the people without any fee. This can promote for the maintenance and expansion of even other indigenous conflict resolution mechanisms.

The *Mezard* can also be used as one of the research sites to attract international researchers and generate an income for the local government. Therefore the *Raya Alamata woreda* have to find a way to promote this precious indigenous culture trough expos and exhibition, and local holidays.

Gender inequality should be maintained. This could be done through enhancing systematic education and administration measures. For instance giving priority and advancing affirmative action, increasing their income by creating job opportunities, etc. as a result when women are becoming leaders in several socio-economic aspects of the community there by their status can be respected in the way that it rearranges the awareness of the members of the community.

Generally, *Mezard* is an institution emanating from the original culture and custom. Thus, preserving and maintaining it is the responsibility of the local government primarily and the regional government generally. Hence, the government should give keen attention to the institution.

Bibliography

Asnake Menbere. (2010). *Inter-Group Conflict in the Awash Valley of Ethiopia : The Case of Afar and Karrayu Oromo*(Ed), in Elbert, F. (2011). *An Anthology of Peace and Security Research Volume 2*, Addis Ababa, Ethiopia.

Assefa Abebe. (2005). *Indigenous Mechanisms for the Prevention of Conflict: The Experience of the Oromo.* Proceedings of an Annual Conference on: Conflict in the Horn: Prevention and Resolution Held In Addis Ababa, June 8-9, OSSREA, Ethiopia.

____(2005) *Luba Basa and Harama Hodha: Traditional Mechanisms For Conflict Resolution In Metekel, West Gojjam* ,Proceedings of An Annual Conference on: Conflict In The Horn: Prevention And Resolution Held In Addis Ababa, June 8-9, OSSREA, Ethiopia.

Bahru Zewde. (2002). *Systems of Local Governance among the Gurage: The Yajoka Qicha and the Gordanna Sera.* Addis Abba: United Printers, Ethiopia.

Barfield, T. and Nojumi, N. (2004). *The Clash Of Two Goods: State And Non-State Dispute Resolution in Afghanistan,* United States Institute of Peace.

Befekadu Zeleke And Diribssa Abate (2005). *Inter-State Ethnic Conflict: Theory and Implications for Ethiopia.* Proceedings of an Annual Conference on: Conflict in the Horn: Prevention and Resolution Held In Addis Ababa, June 8-9, OSSREA, Ethiopia.

Biruk Haile and Jira Mekonnen. (2008). *Customary Dispute Resolution in Harar.* In Alula Pankhurst and Getachew Assafa (eds), *Grassroots Justice In The Contribution of Customary Dispute Resolution.* French Center of Ethiopian Studies, Ethiopia.

Blench, R. and Longtaus, S. et Al. (2006). *The Role of Traditional Rulers in Conflict Prevention and Mediation in Nigeria*, DFID, Nigeria .

Boege, V. (2006) Berghof: *Traditional Approaches to Conflict Transformation Potentials and Limits*, Berghof Research Center.

Brainch, B. (2006). *Commercial ADR, Dispute Resolution Centre,* Kenya.

Center for Democracy and Governance (1998). Agency of International Development, Washington D.C.

Central Statistical Agency (CSA). (2007). *Population and Housing Census. First Draft Report on Population.* Office of the Population and Housing Census Commission, Addis Ababa.

Creswell, W.J. (2003). *Research Design: Qualitative, Quantitative, and Mixed Methods.* Lincoln: University Of Nebraska.

Dahal, R. and Bhatta, C. (2008). *Conflict and Peace Building in the New World Order', Third World Quarterly 20:1*, Washington University in St. Louis.

________________ (2008). *The Relevance of Conflict Resolution Mechanisms for Systematic Conflict Transformation in Nepal,* Washington University,St. Louise,USA.

Dejene Gemechu. (2007). *Conflict and Conflict Resolution among Waliso Oromo of Easter Macha: The Case of the Guma.* Unpublished Master's Thesis, Addis Ababa University, College Of Social Science, Addis Ababa.

Desalegn Chemeda Edossa and Ashim Das Gupta Et Al.(2004). *Indigenous Systems of Conflict Resolution in Oromia, Ethiopia,* Frame Work for Rural Water Management in Africa, Johannesburg, South Africa.

Deutsch, M. and Coleman, P. (2000). The *Handbook of Conflict Resolution: Theory and Practice.* Jossey-Bass, (Eds.).

DFID, Department of International Development Briefing (2004), *Non State Justice and Security Systems.*

Ember, M. (1977). *Anthropology*, New York. Cited in Dejene Gemechu. (2007). *Conflict and Conflict Resolution among Waliso Oromo of Easter Macha: The Case of the Guma.* Social Anthropology dissertation series No. 15

Francis, L. (2006). *An African Overview of Basic Best Conflict Prevention Management and Resolution,* South Africa Peace and Conflict Studies Pretoria University Law.

Goodhand .A and Hulme D. (1999). '*From Wars to Complex Emergencies: Understanding Conflict and Peace Building In The New World Order'*, Third World Quarterly.

Hamdesa Tuso. (2000). *Indigenous Processes of Conflict Resolution in Oromo Society.* In I. William Zartman (Ed), *Traditional Cures For Modern Conflicts*: African Conflict "Medicine", Ethiopia.

Hamer, J. H. (1987). *Dispute Settlement and Changing Norms in Sidamo Associations*, In Ethnology Vol.15.No. 1, Washington DC.

Huyse, L. (2008). *Tradition Based Justice and Reconciliation after Violent Conflicts: Learning from African Experiences*, Brusseles, Belgium.

Ibeanu, O. (2006). *Conceptualizing Peace:* In Shedrack Gaya Best (Ed), *Introduction to Peace and Conflict Studies in West Africa*. Ibadon: Spectrum Books Limited.

IPMS, (2005). *Alamata Pilot Learning Site Diagnosis and Program Design*. Tigray, Ethiopia.

Jeong, H. (2005). *Peace and Conflict Studies, an Introduction*, England: Ashagate Publishing Limited.

Kelemework Tafere (2000). *Indigenous Institution of Conflict Resolution among the Ab'ala Afar of North-Eastern Ethiopia.* Addis Ababa University, College of Social Science, Addis Ababa.

Kennedy, M. H. (2011). *Intra-State Conflict and Indigenous-Based Conflict Resolution Mechanism in Malawi, the Role of Public Affairs Committee*; African Security Review 20.1 26-39, Malawi.

Kothari, R. (2004). *Research Methodology, Methods and Techniques.* New Age International P.Ltd, Publishers, India.

Lowrey, W. (1995). *Sudan Case Study*: *Jinkany-Lou-Nuer Indigenous Peace Process.*

Macfarlane, J. Dr. (2007). *Working Towards Restorative Justice*, Cardozo Journal of Conflict Resolution Vol. 8:487.

Mail H. (2003). *Conflict Transformation*: A Multi Dimensional Task in Bergh of Handbook For Conflict Transformation Http//Www. Berhof. Hand Booknet.Tinfo.

McNamara, C. (2007). "*Basics of Conflict Management*" Adapted From the Field Guide to Leadership and Supervision.

Mellese Madda. (2008). *The Role of Elders In Resolving Conflicts: The Case of Walayta Peoples Of Southern Ethiopia*. In Tarekegn Adebo and Hannah Tsadik (eds), *Making Peace in Ethiopia: Five Cases of Traditional Mechanisms for Conflict Resolution*. Addis Ababa: Peace and Development Committee.

Mensah, F. B. (2000). *Bases of Traditional Conflict Management among the Buems of the Ghana-Togo Borders.*

Meron Zeleke. (2010). Yeshekoch Chilot (The Court of the Sheiks): A Traditional Institution of Conflict Resolution in Oromia Zone of Amhara Regional State. *African Journal of Conflict Resolution, Vol. No.10,* ACCORD, Ethiopia.

Nathan, G. (2007). *The Four Horsemen of the Apocalypse: The Structural Causes of Violent Conflict in Juma (Ed) Conflict Prevention, Management and Resolution In Africa,* Pretoria University Law Press, South Africa.

National Open University of Nigeria (2010). *African Traditional Methods of Conflict Resolution,* Nigeria.

Netsanet Mekonnen (2006). *Abogar-Traditional Conflict Resolution in Huledere Wereda, Ethiopia:* Institute Of Language Studies, Addis Ababa.

Osi, C. (2008). Understanding Indigenous Dispute Resolution Process and Western Alternative Dispute Resolution, *Cardozo Journal of Conflict Resolution Vol.10:183*, Philippines.

Paul, J.L. (1994). *Building Peace and Sustainable Reconciliation in Divided Societie*s, Tokyo:United States University, Japan.

Penal Reform International (2003). *Alternative Dispute Resolution,* L'Exprimeur Publishers, Paris.

Pkalya, R and Adan, M. (2004). *Indigenous Democracy: Traditional Conflict Resolution Mechanisms Pokot, Turkana, Samburu and Marakwet,* Intermediate Technology Development Group, Eastern Africa.

Pospisil, Leopold (1990). "The Attributes of Law". In P. Bohannan(ed.) *Law & Warfare: Studies in the Anthropology of Conflict*. . Austin: University of Texas Press.

Reichel, (1998). *Comparative Criminal Justice Systems*, Prentice – Hall. Bloomfield, New York.

Reilly.L and Bloomfield.T, (1998). *The Changing Nature of Conflict and Conflict Management.* Institute for Democracy And Electrical Assistance, Stockholm.

Robson, C. (2002). *Real World Research.* Blackwell Publishing, London.

Singleton, S. And Shingler, J. (1988). *Africa in Perspective.* USA: Hyden book Company Inc.

Slabbert, D. (2004). Conflict Management Styles in Traditional Organizations. *The Social Science Journal 41 (2004) 83-92,* South Africa.

Spangler, A. (2003). *Conflict Settlement, Resolution, Management and Transformation*: An Explanation Http//Www.Beyaondintractable.Org/Action/Author.Jsp.

Tongeren, P.V.et al. (2005). *People Building Peace II: Successful Stories of Civil Society* Boulder (eds). Lynne Rienner Publishers, Inc.

Tsega Endalew. (2002). *Conflict Resolution through Cultural Tolerance*: *An Analysis of the Micha Institution in Metekel Region*, Ethiopia. Addis Ababa: OSSREA.

United Nations Office on Drugs and Crime (2007). *Training Manual on Alternative Dispute Resolution and Restorative Justice.*

University of Peace, (2005). *A Glossary of Terms and Concepts in Peace Conflict Studies*, Second Edition, Geneva.

USAID, (2005). Sponsored Proceedings On *Practicing Community Based Conflict Mitigation.*

Walter, B. (2000). *Collaboration and Conflict Resolution Skills.* A Core Academic Competence at Camps Conflict Resolution Research Project Vol.1 No.4 State University, Wayne.

Wilson-Fall, W. (2000). *Conflict Prevention and Resolution among the Fulbe.* In I. William Zartman (Ed), *Traditional Cures For Modern Conflicts: African Conflict "Medicine".*

Yacob Arsano, (2006) Sera: *A Traditional Institution of Kambat, in* Bahru Zewde and Siegfried Pausewang (eds), Ethiopian: The Challenge of Democracy from Below. Addis Ababa: United Printers.

Yayneshet Tesfay and Kelemework Tafere. (2004). *Indigenous Rangeland Resources and Conflict Management by the Northern Afar Pastoral Groups in Ethiopia*, DCG Report N0.31, Mekelle, Ethiopia.

Zartman, W.I. (2000). Introduction: *African Traditional Conflict "Medicine". Traditional Cures for Modern Conflicts:* African Conflict "Medicine".

APPENDICES

Appendices A

Questions for interview

The researcher will conduct semi-structured interview with the elders of each "*tabia*" and with the court leaders. The purpose of these interview is to gather a detail and authentic information about the *Mezard* indigenous conflict resolution mechanism, its role in the transformation of conflicts in to peaceful coexistence and also its sustainability and enforceability in rural Alamata woreda, Tigray region, Ethiopia from different opinion leaders/elders, court leaders, and members of the community.

Part one: Questions for the Household Elders

1. What is the meaning and origin of *Mezard* institution of conflict resolution?
2. Who are the main actors involved in resolving conflicts?
3. What are the criteria for selection of the elders?
4. What is the role of the *Mezard* in transforming the conflicts happened in your locality to a peaceful co-existence?
5. What are the stages or procedures held by the *Mezard* indigenous conflict resolution mechanism in terms of for instance; physical damage, murder case, rape and others?
6. What is the mandate of the members of the decision making body in the *Mezard* ?
7. What are the enforcement mechanisms for the decisions passed by elders?
8. How many actors are participated in resolving conflicts?
9. Does the *Mezard* resolve all types of conflict manifested in the community?
10. If it doesn't which types of cases are settled by the *Mezard?*
11. What is the composition of the Jury in the *Mezard?*
    - The criteria for membership?
    - Size of the Jury?
    - Election criteria and opportunity?
12. How can you explain the preference of the people?
13. Does the *Mezard* indigenous conflict resolution mechanism consider the gender composition?
14. What changes has the *Mezard* undergone over the course of time?
15. What is the preference of the people? Why?
16. Are the decisions passed by the *Mezard Jury* binding up on the disputants?

Part two: Questions for the Court Leaders

1. What is the contribution of *Mezard indigenous* conflict resolution mechanism in enhancing Justice?
2. Are there any supporting courses e.g. training given/provided to the male elders?
3. What are the sources of conflict in your woreda and what conflict prevention institutions are practiced to settle the conflicts?
4. Does the formal court recognize any of the decisions passed by the elders on different cases? Which ones? Why? or Why not?
5. Is there any kind of relationship between the formal court and the *Mezard indigenous* conflict resolution mechanism? In what way? Or what is the integration and difference?
6. How do you see the effectiveness of the *Mezard indigenous* conflict resolution mechanism in comparison to the formal court system?
   - Cost
   - Time
   - Proximity
   - Degree of appeal
   - Binding
7. What do you think is the support of the government for the sustainability of the *Mezard* indigenous conflict resolution mechanism?
8. What is the response of the court when there is overlapping of decisions by the formal court system and the *Mezard* indigenous conflict resolution mechanisms?
9. How is the ;
   - Appeal system
   - Acceptance by the people
   - Recognition by law and or country
   - Enforcement mechanisms
   - Binding decisions of the *Mezard?*
10. Any kind of documentation?

    Either in the; charter by itself or in the cases handled.

Part Three: Issues to be raised in the Focus Group Discussion

1. Which mode of conflict resolution do you prefer? The indigenous/*Mezard* or the court? And why?

2. What is the role of the disputants during the process of conflict resolution?

- Obligation
- Swear
- Agreement
- Right
-

3. Does the Mezard institution solve all types of conflict seen with in your community?

4. Are there any mechanisms of the *Mezard* to handle cases in regard to children i.e. below 18 years of age?

5. What is the current status of the *Mezard?*

6. Are there youth participants in the *Mezard* process of conflict resolution?

7. Is there any procedure and mechanisms in the *Mezard* conflict resolution to incorporate female participants in the Jury?

9. What are the weaknesses of the *Mezard* in resolving conflicts?

10. What are the strengths of the *Mezard?*